Colorado State Engineer's Office

Irrigation Laws and Instructions

To Superintendent and Water Commissioners, Colorado

Colorado State Engineer's Office

Irrigation Laws and Instructions
To Superintendent and Water Commissioners, Colorado

ISBN/EAN: 9783337140915

Printed in Europe, USA, Canada, Australia, Japan

Cover: Foto ©berggeist007 / pixelio.de

More available books at **www.hansebooks.com**

Irrigation Laws.

State of Colorado.

1893.

ERRATUM.

Page 134, Section 2227 should be Section 2427.

IRRIGATION LAWS

AND

INSTRUCTIONS

TO

Superintendents and Water Commissioners

COLORADO.

PUBLISHED BY

STATE ENGINEER.

DENVER, COLORADO:
THE SMITH-BROOKS PRINTING CO., STATE PRINTERS
1893

There has been a demand for the laws, governing the supply and use of water for irrigation, in a handy form, for the use of water commissioners and water users.

For the benefit of the officers controlling and the farmers who use the waters of the state, this book was compiled.

I wish to acknowledge the courtesy of secretary of state, N. O. McClees, in publishing.

Respectfully yours,

CHARLES B. CRAMER,
State Engineer.

Denver, August 21, 1893.

LETTER OF INSTRUCTION

TO

Superintendents of Irrigation and Water Commissioners in Colorado.

Issued From the State Engineer's Office,
At Denver, Colorado, July 26, 1893.

Water commissioners will communicate with this office through the superintendents of irrigation of their respective divisions, except in cases of pressing importance or in direct reply to letters from this office.

Superintendents of irrigation will, in a book, designs for which can be seen at this office, enter the tabulated statements provided for in section 7 of Senate bill No. 113, with the additions thereto indicated as desirable below, in columns having the following headings:

Number of division in which situated	Number of district in which situated	NAME OF DITCH, CANAL OR RESERVOIR.	STREAM FROM WHICH WATER IS TAKEN.	Date of appropriation.	Cubic feet of water per sec. decreed to each priority.	Summation of decree to each ditch, canal or reservoir.	Cubic feet of water previously appropriated in district.	Order of priority in district.	Cubic feet of water previously appropriated in division.	Order of priority in division.	Embodied in decree recorded on page...	Rated as entered on page...

Each will be furnished upon application to the state engineer with a book for each district in his division, which book shall be entitled "Register of District No. ____"

Each must enter in said book a tabulated statement relative to the ditches and reservoirs of the appropriate district, which statement shall show in separate columns (as herein below indicated) the

NAME OF DITCH OR CANAL.	Name of stream from which water is diverted.	Order of priority.	DATE OF APPROPRIATION.	Cubic feet of water per second appropriated to each priority.	Summation of appropriations to each ditch or canal.	Cubic feet per second previously appropriated.

And for reservoir priorities the

NAME OF RESERVOIR.	Name of stream from which water is taken.	Order of priority.	DATE OF APPROPRIATION.	Capacity of reservoir in cubic feet.	Cubic feet of water per second appropriated to each priority.	Summation of appropriations to each reservoir.	Cubic feet of water previously appropriated.

They will also certify to the correctness of this tabulated statement.

There shall also be entered in said *register* by the superintendent of irrigation, the rating of each measuring flume in the district, constructed in compliance with section 1813 of the General Statutes. The latter information will be furnished upon application to the state engineer. The *registers* so prepared shall be loaned to the respective water commissioners of his division during the irrigating season, except when occasion demands that they shall be called in for posting.

Said *registers* shall be kept as nearly as possible posted to date by the superintendent of irrigation. They shall be filed, when not needed by the water commissioners, in the office of the state engineer. The superintendents of irrigation shall furnish the water com-, missioners with written instructions to distribute the water of their districts in accordance with the statements in said register.

Superintendents of irrigation will report in writing to this office on the first day of each month, or as soon thereafter as possible, from April to November inclusive. Such report will be accompanied by the reports received from the water commissioners during the previous month. It shall contain such information as the superintendent regards as likely to be of service to the state engineer in his duties, and such suggestions as he may have to make for the advancement of irrigation. It shall especially contain a list of the measuring flumes erected but not rated, and of the measuring flumes previously rated but out of order in each district in his division. Said *registers* shall be delivered to the water commissioners by the superintendents of irrigation by July 1, of each year, if not delivered previous to that date.

Each superintendent of irrigation will be supplied, on application to the state engineer, with blank *Artesian Well Statements*, which he will present either in person or through his water commissioners to the contractor or owner of each artesian well in his division. And he will endeavor to have a statement of each artesian well not already filed in his division filed in the state engineer's office on or before August 15, of each year.

He will have prepared by his water commissioners, for each district in his division, a statement which shall show with reference to each ditch in each district, for each year, in separate columns (as herein below indicated) the

Name of Ditch.	The length thereof in miles.	The number of days water was carried therein during the season of......	The average amount of water in cubic feet, per second, so carried.	Number acres irrigated.	Cost of operation.	Cost of Superintendence.	Cost of Repairs.

And he will file these statements in the state engineer's office by September 30, of each year.

Each superintendent of irrigation will, in person, or through his water commissioners, cause to be made out with reference to each district in his division, and file with the state engineer, on or before August 15, a statement, which shall show with reference to each ditch in each district for the irrigating season, (as herein below indicated) the

Name of Ditch.	Number of acres that can be irrigated therefrom.	Number of acres of alfalfa irrigated therefrom.	Number of acres of seeded grasses irrigated therefrom.	Number of acres of natural grasses irrigated therefrom.	Number of acres of other crops irrigated therefrom.

By seeded grasses is meant cultivated grasses, other than alfalfa, as timothy, clover, etc.

Each water commissioner will embody in a final report to the superintendent of irrigation of the division embracing his district, a full account of his labors during his encumbency of the office of water commissioner, and such final report will be sent to the superintendent of irrigation by October 1.

Each superintendent of irrigation will embody in his final report to the state engineer a full account of the labors of his office during his occupancy thereof, and will accompany his final report with the reports of the water commissioners of his division, and file it with the state engineer on or before October 15. The labors thus put upon superintendents of irrigation and water commissioners are considerable. The allowance for such services, especially that to water commissioners, is small. But the demand for such information is urgent, and the collection thereof can no longer be delayed. The reports of the superintendents of irrigation and those of water commissioners will be embodied to a considerable extent in the published report made by the state engineer to the governor, and it will therein be evident to what extent the different superintendents and commissioners have endeavored to set forth the irrigation condition of their divisions and districts. The plats should be very carefully prepared, as they also will be published as part of the report of the official preparing them.

Water commissioners will notify the state engineer if there is any disagreement as to the *amount* of the bond required of them, upon the part of the several counties into which their **water** district may extend. And in event of such disagreement, will notify the state engineer of the amount of the bond fixed by the various boards of county commissioners of those counties. Thereupon the governor will fix the amount of the bond required, and the water commissioner will be duly notified thereof, and file with the secretary of state.

Water commissioners will notify the state engineer of the approval and filing of their bonds. Superintendents of irrigation and water commissioners will acknowledge the receipt of these *instructions*. .

It is the duty of the water commissioners to be actively employed on the line of the streams in his water district. He should keep himself posted daily as to the flow of water in the streams and as to what ditches are taking water, in order that report thereof may be made at any time on short notice from the superintendent of division.

Locks should be ordered placed on all head-gates where the owners refuse or are unable to keep them closed in accordance with instructions of water commissioners.

Wherever practicable, you will see that waters supplied to ditches in accordance with priority, are beneficially and economically used, or turned back into the streams for the benefit of others.

WATER FOR DOMESTIC USE.

AN ACT

IN RELATON TO "WATER FOR DOMESTIC PURPOSES," PASSED BY
THE EIGHTH GENERAL ASSEMBLY.

Be it enacted by the General Assembly of the State of Colorado:

"SECTION 1. Water claimed and appropriated for domestic purposes shall not be employed or used for irrigation or for application to land or plants in any manner to any extent whatever; *Provided*, That the provisions of this section shall not prohibit any citizen or town or corporation, organized solely for the purpose of supplying water to the inhabitants of such city or town, from supplying water thereto for sprinkling streets and extinguishing fires or for household purposes.

"SEC. 2. Any person claiming the right to divert water for domestic purposes from any natural stream who shall apply or knowingly permit the water so diverted to be applied for other than domestic purposes, to the injury of any other person entitled to use such water for irrigation, shall be deemed guilty of a misdemeanor, and, upon conviction, shall pay a fine of not less than $50 and not exceeding $200, in the discretion of the court wherein conviction is had. Each day of such improper application of water obtained in the manner aforesaid shall be deemed a separate offense. Justices of the peace in their several precincts shall have jurisdiction of the aforesaid offense, subject to the right of appeal, as in cases of assault and battery."

There being no statutory provisions of law governing the distribution of water for domestic purposes, it will not be permitted to carry water in ditches exclusively for such purposes, outside of the order of priorities as established by judicial decrees for irrigation, where such carriage will injuriously affect parties having older rights for irrigation, unless a special order of the court is obtained for such diversion. Further than the above, no laws were enacted by the Eighth General Assembly pertaining to irrigation or affecting the duties of the officers of this department.

The very limited assistants' fund provided for this department by the last General Assembly will render it impracticable to do any rating of ditches where traveling expenses are involved, except as such expenses are paid by the owners of ditches to be rated.

Water commissioners will, therefore, be compelled to exercise their best judgment in the measurement and distribution of water to ditches under decrees, until such time as a fund can be provided for the proper ratings.

It is especially desired that water commissioners, in the exercise of their duties, will collect from all available sources as complete and accurate data as possible relative to the number of acres of land that can be irrigated from each ditch, the number of acres in each of alfalfa, seeded grasses, natural grasses and fruit trees; also, the acreage in all other crops combined.

Pay particular attention to number of acres irrigated from each ditch or reservoir, and the average amount of water flowing in the ditch.

Also ascertain the annual cost of superintendence and repairs for each ditch in your district. Do not confound cost of building or enlarging with cost of running and operating, nor, on the other hand, operating expense with building expense.

Where you can do so, get the cost of building or constructing. In getting, observe the caution given above.

The information thus obtained will be published in the report from this department, and will be of value to each county, as showing the variety and extent of productions therein.

Blanks will be furnished to water commissioners on application to the state engineer for the collection of information regarding the existing reservoirs and reservoir sites—a sample of which will accompany this circular—and, where lands are irrigated by stored waters, it is important to know the number of acres under each reservoir.

In artesian well districts blanks will also be furnished for statements in relation thereto.

For the more accurate and convenient measurement of waters appropriated, pursuant to any decree rendered by any court establishing the claims of priority of any ditch, the owners thereof are required to construct and maintain, under the supervision of the state engineer, a measuring device for measuring the flow of water in the ditch in cubic feet per second.

This measuring device should be an open flume, with apron and wings, constructed as shown on plates "A" and " B " accompanying this letter.

Where the bottom of the ditch exceeds six feet in width the flume is to be sixteen feet in length, and where less than six feet in width the flume is to be twelve feet in length, exclusive in each case of the apron and wings. The width of the flume is to be a little less than the average width of the ditch and a little greater than the width of the ditch on the bottom. The sides of the flume are to be perpendicular, and boarded upon the inside of the posts, and to be of sufficient height to carry the greatest amount of water likely to flow in the ditch. The top of the floor of the flume must be of the grade of the bottom of the ditch. The flume is to be erected on as straight a portion of the ditch as practicable, and not less than 200 nor more than 800 feet below the headgate.

At points 200 feet above and below the location of rating flume it is required that timbers, not less than three inches thick and eight inches wide, and of a length not less than the width of the ditch, be sunk in the bed of the ditch at right angles to the line thereof, until the top of the timber shall be on the grade of the ditch. After the measuring flume is erected, it will be rated by the state engineer, or his assistant, and a datum mark will be made thereon indicating the height to which the water commissioners may raise the water in the flume to allot the ditch the amount decreed thereto by district court. Dead water in the flume should not exceed one-tenth of a foot in depth when the headgate of the ditch is closed.

WATER COMMISSIONERS

Will notify this office of the names of the ditches in your district in which there have been constructed no rating flumes, or in which the rating flumes are out of repair; also of the names and addresses of the managers of such ditches.

C. B. CRAMER,
State Engineer.

U. S. Rev. Stat., 2339. Vested rights—right of of way. "Whenever, by priority of possession, rights to the use of water for mining, agricultural, manufacturing or other purposes, have vested and accrued, and the same are recognized and acknowledged by the local customs, laws and decisions of the courts, the possessors and owners of said vested rights shall be maintained and protected in the same, and the right of way for the construction of ditches and canals for the purposes herein specified is acknowledged and confirmed."

In 1870 Congress further enacted:

U. S. Rev. Stat., 2340, Patents, subject to vested rights. "All patents granted, or pre-emptions, or homesteads allowed, shall be subject to any vested and accrued water-rights, or rights to ditches and reservoirs used in connection with such water-rights, as may have been acquired under or recognized by the preceding section."

These acts have been held to be but the legislative recognition of a pre-existing right, and not the granting of a new right dating from the passage of the law, the formulating of a rule of construction which the courts would have applied without the passage of these laws; Broder v. Natoma Water Co., 11 Otto (U. S.) 274; Osgood v. Eldorado Water Co., 56 Cal., 571; Coffin v Left-Hand Ditch Co., 6 Colo., 446.

CONSTITUTION.

Constitution, article 2, section 14:

Section 287. Taking private property for private use in what cases. That private property shall not be taken for private use unless by consent of the owner, except for private ways of necessity, and except for reservoirs, drains, flumes or ditches on or across the lands of others, for agricultural, mining, milling, domestic or sanitary purposes.

5. See Mills' Ann. Stat. 1890, chap. 45, "Eminent Domain;" see next section and notes; also, see art. 16, sec. 7.

6. Where a person in Colorado, without initiating any steps under pre-emption or other laws to procure title to public lands, places improvements thereon, and another desires to construct his irrigating ditch over or across such lands, if, by a proper proceeding, full compensation is determined, and is paid, for all damages or injury to the improvements caused by constructing such ditch, the constitution and statutory requirements are complied with. Knoth v. Barclay, 8 Colo., 303 (1885).

7. Under these circumstances, the occupant can have no compensation for the taking of the land itself or for injury to the land not taken. Id., 304.

9. Under this section, the taking of private property for private use without the owner's consent must be confined to the purposes named. People ex. rel. v. District Court, 11 Colo., 155 (1887).

10. A tramway for private use is not permitted to be taken, as it is not one of the purposes named in this section. Id.

11. Mills' Ann. Stat., 1890, sec. 3158, is repealed as far as it is inconsistent with this section of the constitution. It was valid as a territorial law, but schedule section 1 of constitution rendered it *pro tanto*, invalid as a state law; Id. Private ways of necessity to haul ore may be taken, but that does not include private railways. Id., 156.

Section 510. **Water public** property. **Sec.** 5. The water of every natural **stream**, not heretofore appropriated, within the state of Colorado, is **hereby** declared to be the property of the public, and the **same** is dedicated **to the use** of the people of the state, **subject** to appropriation, **as** hereinafter provided.

Section 510. Mills' Ann. Stat. 1890, ch. 69, "Irrigation;" **also Colo. Const., art.** 2, sec. 14, **and notes.**

22. **Irrigation defined.** The word **irrigation, in** its **primary** sense, **means** "a sprinkling **or watering,"** but **the best** lexicographers give it **an agricultural or** special signification; thus: "The watering **of lands by** drains or **channels."**—*Worcester.* "The **operation of causing water to flow over lands for nourishing plants."** —*Webster.* **Platte Water Co. vs. Northern Colo. Irr. Co.,** 12 **Colo.,** 529 (1889).

23. **The term irrigation, as** used in Colorado in **the** constitution and **statutes and** judicial opinions, **in view of** the climate **and soil, is in** its special sense **to wit: "The** application **of water to** lands **for** raising **of agricultural** crops and other **products of the** soil." **Id.**

24. The question of water rights **is second to none** in the state in its importance and intricacy, **and a legislative** inquiry under Const., **art. 6, sec. 3, as amended,** cannot require the wholesale **opinion of the supreme court on** secs. **5, 6, 7 and 8 of this article as to irrigation** in **general.** *In re* **S. R. on Irrigation, 9 Col., 620** (1886).

25. **All unappropriated** water **in the natural** streams **of the state is** dedicated **"to the use of the** people," and **the ownership is vested in "the public."** Wheeler vs. **North Colo. Irr. Co., 10 Colo., 587 (1887).**

26. This **section guarantees in the strongest terms** the right of **diversion and appropriation for beneficial uses.** Id.

27. The title to water, after appropriation (except, **perhaps,** the limited quantity actually flowing in the **consumer's** ditch or lateral), remains in the general public, **but** the paramount right to its use, unless forfeited, **continues** in the appropriation. Id., 583.

Section 511. Diverting unappropriated water—
Priority. Sec. 6. The right to divert unappropriated
waters of any natural stream for beneficial uses shall
never be denied. Priority of appropriation shall give
the better right as between those using the water for the
same purpose; but when the waters of any natural stream
are not sufficient for the service of all those desiring the
use of the same, those using the water for domestic purposes shall have the preference over those claiming for
any other purpose, and those using the water for agricultural purposes shall have the preference over those
using the same for manufacturing purposes.

I. COLORADO DECISIONS AND CITATIONS.

(a.) PRIORITY OF APPROPRIATIONS.

1. The common law doctrine as to riparian rights
by which the riparian owner is entitled to the flow of
water in its natural channel upon and over his lands,
whether he makes any beneficial use of it or not, is in-
applicable to Colorado; Coffin v. Left-Hand Ditch Co.,
6 Colo., 447 (1882).

2. In the absence of express statutes to the contrary the first appropriator of water from a natural
stream for a beneficial purpose, has, with the qualifica-
tions contained in the constitution a prior right thereto,
to the extent of such appropriation; Id. Golden Canal
Co. v. Bright, 8 Colo., 148; Hammond v. Rose, 11 Id.,
526 (1888).

3. The doctrine of priority of right to water, by
priority of appropriation thereof for a beneficial purpose, with the modifications declared in the constitution, is, and always has been in force in this state;
Thomas v. Guirand, 6 Colo., 532 (1883); Shilling v.
Rominger, 4 Id., 103 (1878).

4. Water in the various streams in our climate acquires a value unknown in moister climates, here, in-

stead of being a mere incident to the soil, it rises when appropriated, to the dignity of a distinct usufructuary estate, or right of property; Coffin v. Left-Hand Ditch Co., 6 Colo., 446 (1886); City of Denver v. Bayer, 7 Id., 115 (1883); Rominger v. Squires, 9 Id., 329 (1886). •

5. It is and always has been the duty of the state and national governments to protect the right of water in this country by priority of appropriation; Coffin v. Left-Hand Ditch Co., 6 Colo., 446 (1882).

6. The right itself, and the obligation to protect it, existed prior to legislation on the subject of irrigation. Id.

7. It is as much to be protected after patent of the land over which the natural stream flows as before, when such land is part of the public domain; and it is immaterial whether or not such right is or is not expressly excepted from the grant. Id., Hammond v. Rose, 11 Colo., 525-6 (1888).

8. The true test of the appropriation of water is the successful application thereof to the beneficial use designed, and the method of distributing or carrying the same or making such application is immaterial. Thomas v. Guirand, Colo., 533 (1883). Larimer Co. Res. Co. v. People, 8 Id., 616.

9. Appropriation is the intent to take, accompanied by some open, physical demonstration of the intent, and for some valuable use. McDonald v. Bear River Co., 13 Cal., 220. Larimer Co. Res. Co. v. People, 8 Colo., 616 (1885).

10. The true test of the appropriation of water is the successful application thereof to the beneficial use designed, and the method of diverting or carrying the same or making such application is immaterial. Thomas v. Guirand, 6 Colo., 533 (1883). Hayt J. in Farmers' H. L. Canal & R. Co. v. Southworth, 21 Pac. Rep., 1028 (June 17, 1889), 13 Colo., 114.

DAM IN NATURAL STREAM—"DIVERT" AND "APPRO-
PRIATION."

(*b*) 11.   The word "divert" in this section of the constitution must be construed with the word "appropriation," and while the former may mean "to take or
carry away" the water from the bed or channel of the
stream, still the latter means "to successfully apply the
water to the beneficial use designed;" hence, if without
infringing on the prior rights of others, a dam is built
on the bed of a now navigable stream on the public domain, such an act is not unlawful, *per se*, and to comply
with this section of the constitution the water need not
be immediately "taken or carried away" from the dam,
but simply "applied to the beneficial use designed,"
with "reasonable diligence" and "without unnecessary
delay."   Larimer Co. Res. Co. v. People, 8 Colo., 616-7
(1885).

12.   In the absence of any written law upon this
subject, a person would have the legal right to construct
his dam in a non-navigable stream upon the public
domain, and thus preserve water for useful purposes, so
long as he did not in any way encroach upon the superior rights or interests of others.   The government alone
could complain, but it is the policy of both federal and
state governments to encourage the storing of water for
useful purposes in arid districts.   There is nothing in
the unwritten law which countenances the interference
by government with this principle.   Id., 615.

13.   One may make a valid appropriation of surplus
water in the manner above, even though an actual diversion from the bed of the stream did not take place till a
subsequent date.   The act of utilizing as a reservoir a
natural depression, which included the bed of the
stream, or which was found at the source thereof, was
not in and of itself unlawful.   Id., 617.

14.   But he who attempts to appropriate water in
this way does so at his peril.   He must see to it that no
legal right of prior appropriators, or of other persons is
in any way interfered with by his acts.   He cannot
lessen the quantity of water, seriously impair its quality,

or impede its natural flow, to the detriment of others who have acquired legal rights therein superior to his. And he must respond in proper actions for all injuries resulting to them by reason of his acts in the premises. Id.

15. While the legislature cannot prohibit the appropriation or diversion of unappropriated water for useful purposes, from natural streams upon the public domain, that body has power to regulate the manner of effecting such appropriation or diversion. It may by reasonable and constitutional legislation, designate how the water shall be turned from the stream, or how it shall be stored and preserved. Id., 618.

WATER IS APPROPRIATED BY USE, NOT GRANT OR CHARTER.

(c) 16. No company, unless it has a charter that will be protected under the federal constitution from any state interference, constitutional or legislative, by which its obligation is impaired, can acquire any rights for the purposes of irrigation of a different kind or superior to such as are acquired by priority of appropriation. Platte Water Co. v. North Colo. Irr. Co., 12 Colo., 530 (1889).

17. There is no local custom or judicial decision of Kansas or Colorado which declares that a party may secure a grant to the exclusive use of the water of a natural stream, allow the same to remain in abeyance for a long series of years without making use of the exclusive privilege so granted, and thereafter assert the same to the exclusion of those who have in the meantime acquired rights to the use of such waters by actual appropriation and use in pursuance of the general laws of the state. Id., 531.

18. By the constitution and laws of Colorado, state and territorial, from the earliest times, rights to the beneficial use of water from natural streams have been acquired by diversion through prior appropriation rather than by grant. It has been the settled doctrine of our courts that such appropriation, to be valid, must be manifested by the successful application of the water

to the beneficial use designed or accompanied by some
open, physical demonstration of intent to take the same
for such use. Id., Yunker v. Nichols, 1 Colo., 555;
Schilling v. Rominger, 4 Id., 103; Coffin v. Ditch Co., 6
Id., 446; Thomas v. Guirand, Id., 532; Sieber v. Frink,
7 Id., 154; Reservoir Co. v. People, 8 Id., 616.

19.   At common law the right to use the running
water of a natural stream, not navigable, is an incident
to the ownership of the soil through which it flows.   It
is a part of the freehold, and passes only by grant or
prescription.   Platte Water Co. v. North Colo. Irr. Co.,
12 Colo., 532 (1889); 3 Kent Com. 439; Ang. Water
Courses. sec. 8; Gardner v. Village of Newburg, 2
Johns, ch. 162; Davis v. Fuller, 12 Vt., 178.

20.   At common law the right to divert the water
of a non-navigable stream might be granted to a certain
extent, by the proprietor of the soil through which the
stream runs.   But in the controversy under consideration such proprietor was the United States, and appellant, (Platte Water Co.), shows no grant from such
proprietor. On the contrary, the act of Congress of May
30, 1854, section 24, under which the territory of Kansas
was organized, in conferring legislative power upon the
territory (and from which territory appellant's charter
came) expressly provided that "No law shall be passed
interfering with the primary disposal of the soil;"
Platte Water Co v. North Colo. Irr. Co., 12 Colo., 532
(1889).

21.   The principal of priority of appropriation
must not be ignored in a decree apportioning water
between consumers; Burnham v. Freeman, 11 Colo.,
605 (1888).

DOCTRINE OF PRIORITY EXISTED PRIOR TO CONSTITU-
TION.

(*d*) 22.   The doctrine of priority of right to water
by priority of appropriation has existed in Colorado
from the date of the earliest appropriation of water, and
it was not first recognized and adopted in the constitution; Coffin v. Left-Hand Ditch Co., 6 Colo., 446 (1882).

22a. The act of congress (14 U. S. Stat. at L. 251, sec 9, approved July 26, 1866, and amended July 9, 1870, 16 U. S. Stat. at L. 218, sec 17; same Mill's Ann. Stat. Appendix) protecting priority of water rights, is a **voluntary** recognition of a pre-existing right of possession constituting a valid claim to its continued use, and not the establishment of a new one; Id., 447; Broder v. Natonia Water & M'g Co., 101 U. S., 276 (1879); City of Denver v. Mullen, 7 Colo., 363 (1884); and see Shilling v. Rominger, 4 Id., 109 (1878); Hammond v. Rose, 11 Id., 526 (1888); Platte Water Co. v. North Colo. Irr. Co., 12 Id., 533 (1889).

23. The act of Congress above mentioned says nothing about the rights claimed by legislative grants without actual possession, appropriation and use; Platte Water Co. v. North Colo. Irr. Co., 12 Colo., 531 (1889).

PRIORITY MAY RELATE BACK TO TIME DITCH WAS BEGUN—DILIGENCE—REASONABLE TIME.

(e.) **24.** Although the appropriation is not deemed complete until the actual diversion or use of the water, still if such work be prosecuted with reasonable diligence, the right relates to the time when the first step was taken to secure it. Ophir M. Co. v. Carpenter, 4 Nev, 544; Kelly v. Natonia W. Co., 6 Id., 109; Sieber v. Frink, 7 Colo., 153 (1883); Larimer Co. Res. Co. v. People, 8 Id., 617 (1885); Wheeler v. North Colo. Irr. Co., 10 Id., 588 (1887).

25. One of the essential elements of a valid appropriation of water is the application thereof to some useful industry. To acquire a right to water from the date of the diversion thereof, one must within a reasonable time employ the same in the business for which the appropriation is made. What shall constitute such reasonable time is a question of fact depending upon the circumstances connected with each particular case. Sieber v. Frink, 7 Colo., 154 (1883); Hayt, J., In Farmers' H. L. C. & R. Co. v. Southworth, 21 Pac. Rep., 1029 (June 17, 1889); 13 Colo., 115.

26. To constitute a legal appropriation, the water must be applied within a reasonable time to some beneficial use, that is the diversion ripens into a valid appropriation only when the water is utilized by the consumer. Id. Platte Water Co. v. N. Colo. Irr. Co., 12 Colo., 531 (1889); Farmers' H. L. Canal & R. Co. v. Southworth, 21 Pac. Rep., 1029, 1030 (June 17, 1889), 13 Colo., 115.

STREAM MAY BE CARRIED INTO A DIFFERENT WATER-SHED—POINT OF DIVERSION.

(*f.*) 27. It is competent to take the water from one stream and carry it over a water-shed to a different drainage, and priority of appropriation will still obtain. Coffin v. Left-Hand Ditch Co., 6 Colo., 449, etc. (1882); Thomas v. Guiraud, Id. 532 (1883).

28. Change in point of diversion does not affect a party's right to priority. Sieber v. Frink, 7 Colo., 154 (1883).

29. The right to priority of appropriation is in no way dependent upon the locus of its application to the beneficial use designed. Hammond v. Rose, 11 Colo., 526 (1888); Coffin v. Left-Hand Ditch Co., 6 Id., 449, (1882.)

30. The lands to be irrigated need not be on the banks of the river, nor in its neighborhood, and may be on a different drainage. Id., Hammond v. Rose, 11 Colo., 526, (1888.)

POINT OF DIVERSION, PLACE AND CHARACTER OF USE, MAY ALL BE CHANGED IF NO ONE IS INJURED.

(*g*) 31. In the case of Sieber vs. Frink, 7 Colo., 154, the use made of the water diverted and the land upon which it was applied remained the same, the only change was some eighty feet in the point of diversion; and no one was injured by this change, and it was held to have been lawful. Fuller v. Swan River P. M. Co., 12 Colo., 16 (1888.)

32. Held, now also that where the right of no one is injured, a change in the point of diversion can be made for the purpose of changing the place of the use. Id., 19.

33. Where the right of no one is injured, the right
to change includes the point of diversion and the place
and character of use. Id.

34. A party who makes a prior appropriation of
water can change the place of its use without losing
that priority as against those whose rights have attached
before the change. Id., 17, Maeris v. Bicknell, 7 Cal.,
263.

35. The rights of an appropriator to the water of
a stream are strictly usufructuary, and in all cases the
effect of the change upon the rights of others is the controlling consideration, and in the absence of injurious
consequences to others, any change which the party
chooses to make is legal and proper. Fuller v. Swan
River P. M. Co., 12 Colo., 19 (1888); Kidd vs. Land, 15
Cal., 162, 180.

36. The right to change the point of diversion is
absolute, except as limited by the condition "that the
change must not injuriously affect the right of others."
Fuller v. Swan River P. M. Co., 12 Colo., 19 (1888).
Mining Co. v. Morgan, 19 Cal., 609, 616.

37. Appropriation, use and non-use are the tests
of an appropriator's right, and place of use and character of use are not. When he has made his appropriation he becomes entitled to the use of the quantity
which he has appropriated at any place where he may
choose to convey it, and for any use and beneficial
purpose to which he may choose to apply it. Any other
rule would lead to endless complications, and must
materially impair the value of water rights and privileges. Fuller v. Swan River P. M. Co., 12 Colo., 17
(1888). Davis v. Gale, 32 Cal., 27.

38. One entitled to divert a quantity of water from
a stream may take the same at any point on the stream,
and may change the point of diversion at pleasure, if
the rights of others be not injuriously affected by the
change. Junkans v. Bergin, 67 Cal., 267-270. Fuller
v. Swan River P. M. Co., 12 Colo., 17 (1888).

(h) Abandonment. See sec. 3138, note 1.

39. A failure to use for a time is competent evidence on the question of abandonment, and if such non-user is continued for an unreasonable period, it may fairly create a presumption of an intention to abandon; but this presumption is not conclusive, and may be overcome by other satisfactory proofs. Sieber v. Frink, 7 Colo., 154 (1883). Instance of abandonment, Dorr v. Hammond, 7 Colo., 83 (1883).

40. It is not reasonable to suppose that priority of right to water, where water is scarce or likely to become so, will be lightly sacrificed or surrendered by its owner, nor should the owner of such a right be held to have surrendered it or merged it, except upon reasonably clear and satisfactory evidence. Rominger v. Squires, 9 Colo., 329 (1886).

41. Where parties with different priorities in water from an old ditch agree to build a new one, but in such agreement say nothing as to the division of the water, it must not be inferred that such silence in the contract is any waiver of priorities, and a court has no authority to reduce their priorities to a common date. Id.; Hayt, J., in Farmers' H. L. Canal & R. Co. v. Southworth, 13 Colo., 114–136; Helm, J., in Id., 118-9. Instance of abandonment: Dorr v. Hammond, 7 Id., 83 (1883).

42. The constitution and statutes recognize the right to construct and maintain reservoirs. (See Mills' Ann. Stat., 1890, sec. 2270.) Larimer Co. Res. Co. v. People, 8 Colo., 615 (1885).

INTERNAL IMPROVEMENT.

(*i*) 43. A system of reservoirs and canals for the purpose of storing and delivering water to all within reach thereof, with the control retained by the state, is an "internal improvement," to which the "internal improvement fund" may be devoted. *In re* S. R. as to Inter. Imp. Fund, 12 Colo., 286-7 (1888); *in re* S. R. as to Inter. Imp. Fund of Enabling Act, Id., 288.

44. The constitution says nothing about changing channels of natural streams, but the right to divert the

unappropriated waters of natural streams to beneficial uses is guaranteed. Id.

45. In any canal or reservoir system devised the constitutional rights of prior appropriators must not be invaded. Id.

REGULATION OF USE OF WATER.

(j) 46. A determination of the priorities of water rights is incidental to a proper regulation of the use of water diverted from the natural streams of the state. Golden Canal Co. v. Bright, 8 Colo., 147 (1884).

47. The payment of expenses and costs in determining such priorities would naturally be considered in an act regulating the use of water. Id., 148.

DITCH DECREES STATUTORY.

(k) 48. The acts of 1879 (L. 79, p. 94, etc.), and 1881 (L. '81, p. 142, etc.); same G. S., '83, pp. 571-584; same (Mills' Ann. Stat. 1890, sec. 2399-2439), as to the settling of priority of rights by referee, etc., provided a purely statutory proceeding to determine the priority of rights to the use of water for irrigation, between ditch, canal and reservoir owners, taking water from same natural stream; this proceeding cannot be used for the purpose of determining the claims of parties to the use of water for domestic or other purposes, not fairly within the term irrigation, as above defined. Platte Water Co. v. N. Colo. Irr. Co., 12 Colo., 529, 533 (1889).

49. The right of the appellant company (Platte Water Co.) as an appropriator of water for purposes other than irrigation, and the rights of the city of Denver (not a party to the record) as a consumer of water for any purpose, cannot be adjudicated by the courts in such proceedings; Id., 534.

(1). The Wheeler case—The pleadings.

50. The case of Wheeler v. The Northern Colorado Irrigation Co., 10 Colo., 582 (Jan. 4, 1888); S. C., 17 Pac. Rep., 487, was a mandamus proceeding. The appellant, Wheeler, as relator, brought the action to

compel the respondent company to furnish him water for the irrigation of his lands, under the ditch of said company. The alternative writ was granted and demurred to by the company, and the demurrer sustained, and from the judgment entered on said demurrer in favor of the respondent, said Wheeler appealed. The material allegations of the petition (and alternative writ, which, with the exception of its introduction and conclusion, is a verbatim copy, of the petition. Printed abstract, fol 52), are as follows :

1.   That appellant is a corporation.

2.   That it was incorporated to appropriate water out of the Platte river, to conduct it over a specified route, and for compensation therefor, to supply it to the tillers of the land lying thereunder, with which to irrigate the same.

3.   That pursuant thereto, it constructed a ditch, and run the water from said stream therein.

4.   That petitioner is the owner of land under said ditch, which is arid, and by reason thereof it cannot be made fruitful without the aid of said water.

5.   That said land is dependent on said water, there being no other source of supply therefor ; and relator relying on his right to his share of the water running in said ditch, not otherwise appropriated, set out and has now growing trees, shrubbery, crops, etc., which will die unless irrigated.

6.   That relator applied for and demanded of respondent water with which to irrigate said growing things, and to that end tendered to it " the price of water established and demanded by respondent as water rates for such purposes;" also offered and was ready, able and willing to conform to and be governed "by all lawful and reasonable rules respecting the use of said water " running in said ditch ; and that said demands, tenders, offers and water were each refused, unless he would sign the contract set out below and agree to pay the royalty exacted, and grant the company right of way through his lands for its ditch without compensation.

7. That **at said** time appellant had running in its said ditch, unappropriated, and not being used for any beneficial purposes, the waters wherewith **to** supply relator.

51. **The** great importance that attaches to this **case** of Wheeler vs. The Northern Colorado Irrigation **Co.**, both justifies **and** requires its fullest presentation. **The** printed reports of this case fail to set out the **contract** involved in the suit, and the opinions of Helm, J., **and** of Beck, Ch. J., contain a quotation of but one **or two** sentences **taken from** the contract. Justice Elbert did not sit **in the case, and** there is a separate opinion by each **of** the other justices. Helm, J., **writing** the principal **opinion, says:** "This **contract** contains a number of **conditions** that appear unreasonable, **and** as I construe the constitution **and statutes are** of doubtful legality. **But** it is sufficient **to recall** the fact that the **unlawful** demand of $10.00 per **acre** for the right to use **water is** a conspicuous provision therein. Relator could no **more** be required to execute a contract containing this condition, than he could be compelled **to** comply with the demand in the absence of contract." Wheeler case, 10 **Colo.**, 596; see also opinion of Beck, Ch. **J.**, Id., 597.

(*m*) Contract involved **in Wheeler case, Id.** *haec verba.*

52. **That** the entire contract as before **the court,** when the **above** was written **by** Justice Helm, may **be** placed at the service of the profession, I have taken from the files of this case (No. 1891) in the office of the clerk of the supreme court, pp. 9–14, fols. 24–39, of the printed abstract, which contains all the allegations in the said petition with reference to the contents of **the** said contract, and they are verbatim as follows, to-wit:

(**Fol.** 24.) Petitioner further represents that said **company**, seeking and intending to wrongfully and unlawfully oppress your petitioner, as well as divers and **sundry** other citizens owning land under said ditch, and **depending** thereon for water for irrigation, has caused **to be drawn** up, printed and written, certain contracts so-called, which it requires your petitioners, and others in like situation with him, to sign as one of the conditions of purchasing the right to use (25) water, which said contract contains requirements and conditions of performance on the part of those so needing and purchasing the use of said water, and envolving certain forfeitures of rights and privileges thereof, of the most exacting, unfair, inequitable and unlawful character, insomuch that those who consent to sign the same, knowing their terms and conditions, are forced to do so as the sole condition upon which they can procure water from defendants canal and cultivate their land. Said contracts, after reciting that, in consideration of the stipulations (26) therein contained, and the payments as therein specified, the said company, party of the first part agrees to sell to the consumer of water, the party of the second part, "the right to receive and use water from the canal of the first party," for irrigating the land described, for the sum of money named, and also "upon the further payment annually, in advance on or before the first day of May in each year from the date hereof, such a reasonable rental per annum, not less than one dollar and a half per acre, and not more than four dollars per acre, as may be established from year to year by the first party," and after setting forth several enumerated (27) rules and regulations, more or less reasonable, respecting the use of water by the consumer thereof, the said second party, said contracts thereafter contain the rules, conditions and stipulations as follows, to-wit:

"*Seventh*—In case of any dispute between the different parties, as among themselves, to the use of water from the main canals, laterals or subsidiary (28) canals or ditches of said first party, the same shall be referred to the superintendent of said first party and his decision shall be final and binding upon all the parties interested.

"*Eighth*—And the said party of the second part, for ----and ---- heirs, and assigns agree---- in consideration aforesaid, to waive and hereby do waive any and all claims for loss or damage, by reason of any leakage, seepage or overflow, from any canals or ditches, or from any reservoirs, lakes or laterals of said first party, either upon the land aforesaid, or any other tract belonging to ____ anything in any statute, law (29) or custom to the contrary, notwithstanding.

"*Ninth*—And the said party of the first part, at all times hereafter, shall have the right to add to, and change and modify, the foregoing rules and regulations, or any of them, so far as may be reasonably necessary to regulate the delivery and distribution of water to said party of the second part, ____ heirs or assigns.

"And it is hereby agreed and covenanted by the parties hereto, that time and punctuality are material and essential ingredients of this contract, and in case the second party, ____ heirs (30) or assigns, shall fail to make the payments aforesaid and each of them punctually, and upon the strict terms and times above limited, or shall fail, neglect and refuse to take and pay rent for said water in accordance with the contract for any two years in succession, or shall fail to perform and complete all and each of said agreements and stipulations aforesaid strictly and literally, without any failures or defaults, then this contract, so far as it may bind the first party, shall become utterly (31) null and void, and all rights and interests hereby created or then existing in favor of the second party, or derived from ____, shall utterly cease and determine, and all equitable and legal interests in the premises, hereby contracted to be conveyed, shall revert to and revest in said first party without any declaration of forfeiture or act of re-entry, or any other act of said first party to be performed, and without any right of said second party of reclamation or compensation for moneys paid, as absolutely, fully and perfectly as if this contract had never been made.

(32) "In case the second party, ____ legal representatives or assigns, shall pay the several sums aforesaid punctually and at the time above limited, then the said party of the first part shall (upon request at the general office of the first part and the surrender of this contract) execute and deliver unto the said party of the second part, ____ heirs or assigns, a deed conveying

the right to use the water herein agreed to be sold, subject, however, to all of the conditions, restrictions and exceptions hereinbefore mentioned.

"And the said party of the first part shall have the right immediately, or at any time thereafter, upon the failure of the party of the second part to (33) comply with the stipulations of his contract, to refuse and to cease to supply any water under this agreement. And the said party of the second part covenants and agrees that ____ will surrender unto the said party of the first part all rights or interests hereby created without delay or hindrance, and no court shall relieve the party of the second part, ____ heirs or assigns, from the failure to comply strictly and literally with this contract.

(34) "And it is further stipulated that no assignment of this contract, or sub-sale of the premises, shall be valid or binding on the said party, nor shall said first party be bound or affected by any notice, actual or constructive, by record or otherwise, of any such assignment or sub-sale, unless the same shall be made by indorsement thereon or permanently attached thereto, and this contract, so assigned, be presented or sent to the first party, and the assignment or sub-sale approved by the first party; and the conveyance of said premises by the first party to said purchaser, or, in case of assignment (35) or sub-sale, as aforesaid, to the assignee named in the last assignment so approved, shall be deemed to be a full and complete performance of this contract against all persons claiming any title to or interest in said premises, under or by virtue of this contract, and it shall be within the power of the first party to approve, or refuse to approve, any assignment or sub-sale of the premises."

Petitioner further represents that in addition to all the said rules and conditions and the water rates fixed and demanded, as aforesaid, the said company (36) requires as a condition precedent to the purchase and use of said water by petitioner and others in like situation that they shall pay to said company as a bonus or gift a certain price per acre, to wit: The sum of from ten to thirty dollars per acre for each acre of land upon which water is to be furnished and used, for the privilege of buying and using said water; that defendant gives no equivalent or consideration for such bonus or gift so demanded, and petitioner represents that this so-called bonus or gift is in effect intended to be simply an invasion of the law of the land and the rights of the water (37) consumers, in that it is but an indirect mode of increasing and adding to the lawful and reasonable price of the water as otherwise fixed by said company, at the rate of a certain sum per cubic foot, and paid for at such established rates by all consumers, forasmuch as it is not pretended that such bonus or gift demanded is the price of the water sold and consumed at fixed rates, as aforesaid, nor the price of the land cultivated, nor rental of lands, nor repairs of the ditch, nor for any other specific article, thing, use or (38) rightful purpose whatever, not otherwise paid for; but petitioner avers that in addition to all other requirements, exactions, payments and conditions, rightful and wrongful, demanded of consumers of water by said company, this so-called bonus or gift is an arbitrary, extortionate, unjust and unlawful exaction for a right and privilege already existing and otherwise fully paid and compensated for.

Petitioner states further and admits that he has hitherto refused and still refuses to pay the amount demanded as a bonus or gift aforesaid, for the reason that he believes and is so advised that the demand and exaction of such bonus or gift is unlawful, unreasonable and without any right whatever on the part of said company in the premises.

(39) Petitioner states that the sum demanded of him by said company as said bonus or gift is the sum of $10 per acre, which amounts to over $2,200 for the amount of land, to wit: over 220¼ acres, which he desires to cultivate, and has applied for water as aforesaid.

WATER RIGHTS—IMPORTANCE.

(n) 53. Water rights have always been justly regarded as one of the most important subjects dealt with in the legislation and jurisprudence of Colorado. Wheeler v. North Colorado Irrigation Co., 10 Colorado, 586 (1888).

53 (*a*). Adjustment of priorities and differences of consumers has hitherto received chief attention, but thereafter the status of the carrier and its relations with the consumers will command most earnest and thoughtful consideration. Id.

CARRIERS, CONSUMERS, CO-CONSUMERS.

(*o*) 54. The terms "carrier" and "consumer" are used as meaning the canal company and tiller of the soil, respectively, Id.; Elliott, J., in Farmers' H. L. Canal & R. Co. v. Southworth, 21 Pac. Rep., 1030 (June 17, 1889); 13 Colo., 132; Helm, J., in Id, 119; and in this last case Helm, J., uses the term "co-consumer" to designate those consumers taking water from the same artificial stream.

55. Colorado farmers, with few exceptions, are unable to convey water from the natural streams to their land; Wheeler v. North Colorado Irr. Co., 10 Colo., 586 (1888).

56. To successful cultivation in the arid region, the carrier and consumer are equally indispensible ; and the rights of both are to be protected by the courts; Id.

CANAL COMPANY IS COMMON CARRIER, NOT PROPRIETOR OF WATER.

(*p*) 57 The status given the carrier of water (canal company) is exceptional—differing in some particulars from that of an ordinary common carrier. Certain peculiar rights are acquired in connection with the water diverted, which are dependent for their birth and continued existence upon the use made by the consumer ; Id. 588; Elliott, J., in Farmers' H. L. C. & R. Co. v. Southworth, 21 Pac. Rep., 1030; 13 Colo., 131, (June 17, 1889.)

58. But, giving these rights all due significance, the carrier cannot become the "proprietor" of the water

diverted. **Wheeler v. North Colo. Irr. Co.,** 10 Colo., 588 (1888).

CANAL COMPANY IS QUASI PUBLIC AGENT.

(*q*) 59. The carrier (canal company) under the constitution is a *quasi* public servant **or** agent; Wheeler **v. North Colo. Irr. Co.,** 10 Colo., 588 (1888).

60. **It is** permitted to acquire certain **rights as** against those subsequently diverting water **from the** same natural stream. It·may exercise the **right of** eminent domain, but **it is** charged with **certain duties** and subject to **reasonable** control; Id.

61. **The carrier (canal company)** voluntarily engages **in irrigation enterprises, and in most** instances, from the nature **of** things **has a** monopoly of the business **ness** along the line of its **canal;** Id.

CANAL COMPANY CHARGED WITH A PUBLIC TRUST.

(*r*) 62. The vocation of the canal company, and the ' **use of its** property are closely allied to the public interest; **its** conduct in connection therewith materially affects **the** community at large—it is charged **with** what the decisions term a public duty or **trust;** Id., **590.**

63. Water rates must **be reasonable—may be** regulated. For these reasons by **the** common **law, even if** the constitution and statutes were absolutely **silent on** the question of charges for transportation, **and the time** and manner of their collection, it must **be held** that the canal company **has** submitted itself to **a** reasonable judicial control, invoked and exercised for the common good in **the matter** of regulations and charges, **and that** such charges **must be** reasonable; Id., 589-90.

64. **If a canal** company **(carrier)** attempts to **use** its monopoly **for** the purpose of coercing compliance **with** unreasonable and exorbitant demands, it invites judicial interference; Id., 590; Munn v. People, 94 U. S., **113;** Price v. Riverside L. L. Co., 56 Cal., 431; C. & N. **W. R. R. Co. v. People,** 56 Ill., 365; Vincent v. Chi. & **Alton R. R. Co.,** 49 Id., 33.

65.   The carrier must be regarded as an interme-
diate agency, existing for the purpose of aiding consumers in the exercise of their constitutional rights, as well
as private enterprises, prosecuted for the benefit of its
owners; Wheeler v. North Colo. Irr. Co., 10 Colo., 590
(1888).

66.   According to the literal terms of the contract
calling for $1.50 annually and $10.00 in addition (fols.
26, 36 supra, note 52), the $10.00 exaction is illegal, Id.,
591.

67.   Construction claimed for contract.   It is said
the $10.00 per acre is not for the right to use water, but
is merely a portion of the annual "rental" exacted from
customers in advance for the remaining ten years of the
canal company's existence; that instead of requiring,
say, $2.50 per acre for each irrigating season in turn, the
canal company has seen fit to divide this sum into two
parts, collecting $1.50 annually, and the residue of $1.00
each for the remaining ten years of its corporate life,
as one entire sum in advance; Id.

68.   This construction of the contract is plausible,
but it is doubted if the courts could accept it.   Id.

SUCH A CONSTRUCTION WOULD DENY CONSUMERS CON-
STITUTIONAL RIGHTS.

(s)   69.   But, if accepted the position is not tenable.
If in the absence of legislation the carrier can charge
for part, it can charge for all its annual transportation
charges in advance ; and if the company's life has
twenty years to run, it can compel payment for the cost
of delivering water for the whole twenty years before
the consumer can exercise his constitutional right during
a single season, and he may not want the water for
twenty years, and may not be able to advance so large a
sum at once.   To say consumers must do so or have no
water, is to deprive them of their constitutional right as
effectually as though the right itself had no existence.
Wheeler v. North Colo. Irr. Co., 10 Colo., 591-2, (1888.)

70.   If this position were allowed, the consumer's
right in section 6, article 16 would, in the absence of

legislation, be subject for its efficacy to the greed or caprice of a single individual or corporation. Id., 592.

SUCH CONSTRUCTION UNJUST—OPPRESSIVE—NO GUAR-
ANTY THAT COMPANY WILL LIVE ITS
STATUTORY LIFE.

(*t*) 71. Such position, if allowed too, would consummate a most unreasonable and unjust discrimination. Id.

72. The consumers who pay for twenty years in advance, have no assurance that the carrier will keep its engagement during that period. Id., 594.

73. The said demand of $10 per acre as an advance payment of part of the transportation charge, for the remaining years of its corporate life is illegal, unreasonable and oppressive. Id., Beck, C. J., Id., 597, 599.

TIME AND CONDITION OF PAYMENT TO BE FIXED BY
LEGISLATION—COUNTY COMMISSIONERS
FIX AMOUNT.

(*u*) 74. Against such exaction, the consumer's only remedy is not by application to the county commissioners; they can fix the maximum amount, but not the time or conditions of payment. Id., 593.

75. Time and conditions are proper subjects for legislation. The legislature might provide that the maximum rate fixed by county commissioners, or such lower rate as the carrier might charge, should be paid annually in advance for such irrigating season, or it might make any other reasonable regulations in these respects; but neither the legislature nor the carrier could make regulations or rules by which the canal company could accomplish a wholesale discrimination between consumers, and deny to a majority of them, if it chooses, the rights secured by the constitution. Id.

76. Had there been a statute providing a method of procedure before the county commissioners, where the head of a ditch was in another county than that in which plaintiff resides, he would have been compelled

to have first applied to said commissioners, before he could resort to the courts, but there was no such statute. Id., 599.

TEN DOLLARS ROYALTY—VOID, BUT NOT PER SE. PRIOR TO ANTI-ROYALTY ACT.

(*v*) 77. No expenditure in building ditches, however vast, and no inconvenience, however great, can justify or legalize the said exaction of $10.00 per acre as an absolute condition precedent to use water for the current irrigating season.   Wheeler v. North Colo. Irr. Co., 10 Colo., 595 (1888).

78. But the consumer must object, as said exaction is not illegal, *per se*.   Id.

79. If the consumer, prior to 1887 (the anti-royalty act was not passed until April 4, 1887, L. '87, p. 308; same, Mills' Ann. Stat., 1890, sec. 2304), voluntarily submitted to such exactions, both the legislature and the courts may be powerless to relieve him from the legitimate results of his contracts; Id.   But Beck, C. J., says, "Any sum charged for royalty as a bonus would be unconstitutional."   Id., 599.

30. A large portion of said $10.00 may have been wholly for royalty, gift or bonus, but the record does not warrant the proposition that it all was, and that the $1.50 was alone the full tranportation charge.   Beck, C. J., Id., 597.

CONFUSING EXPRESSIONS.

(*w*) 81. The words in the contract "the right to use water," etc. (fol. 26), are open to criticism, as appearing to sell a right, that is by the constitution dedicated to the people and vested in the public, and, therefore, not a subject of sale; but the constitution in article 16, section 8, speaks of fixing rates "for the use of water" and the statutes contain the expressions "selling water," furnishing water for sale," "purchasing water," etc., Id. 598.

(*x*) 82. The Southworth case· The case of the Farmers' High Line Canal & Reservoir Co. *et al.* v.

Southworth, 21 **Pac.** Rep., 1028, **13 Colo.,** 111 (1889), was a suit for injunction brought by Southworth to enjoin the said company, and certain consumers whose priorities were alleged to be junior to his, from prorating (on account of alleged scarcity) the water claimed **by** plaintiff under right of priority, pursuant to the prorating statute (Mills' Ann. Stat., 1890, sec. 2267) Elliott, then judge of the district court of Arapahoe **county,** overruled the demurrer filed by the defendant **and held** the complaint stated **a good** cause of action. Defendant appealed **to** the supreme court; separate **opinions** were filed by **each of the judges,** among whom **was now** Judge Elliott, **who heard** the case below. The **constitutional** questions **involved in** this **action are so vital and** serious that I shall **endeavor to show the present status of the law,** by setting **forth the separate** holdings **of each of the** justices.

PLEADING PRIORITIES.

83. **Held,** by Hayt, J., **that the** complaint **was insufficient** because while it **alleged the** priority **of** right in plaintiff **to** the appropriation of water, such allegation was **a** conclussion of law, and the facts of diversion and appropriation to a beneficial use, which constitutes such priority, should have **been** specifically alleged. Farmers' High Line Canal & Reservoir **Co. v.** Southworth, **13** Colo., **115** (1889).

84. Elliott, J., **also holds** the same **(Id., 130) and** that the complaint should have alleged that **plaintiff** was **"accustomed to take and** apply the **water** without waiver **or abandonment to his** crops and **trees."** Id. 139. He gives an **ellaborate discussion** on the merits, however, regardless **of the alleged defects in** the complaint.

85. Helm, C. J., denies the **foregoing propositions** in the following language: "The **complaint states** certain conclusions of law, **and might have been more** artificially drawn in other respects; **but, after eliminating** these legal conclusions, the following **alleged facts** may, I think, be fairly gathered from the **remaining** averments, viz: That defendant, the High **Line Company,** is a corporation duly organized under **the laws of**

the state, and is doing business as a carrier of water; that plaintiff is a consumer on the line of defendant's canal; that on or about the first of April, 1881, plaintiff procured water through the defendant's canal to irrigate his land, which use has not been abandoned; that plaintiff has paid and defendant has accepted the charge for transporting to him during the season of 1887 the quantity of water he has previously used; but that defendant, there being a probable scarcity, threatens to pro-rate, and has taken steps so to do, the diminished quantity to which the canal will be entitled, between plaintiff and certain consumers who began taking from defendant's canal subsequent to the said first day of April, 1881. The object of the action is to enjoin such pro-rating, and compel defendant to allow plaintiff the entire quantity heretofore used by him, regardless of the interests of those co-consumers whose use post-dates that of plaintiff, and regardless of the command embodied in the pro-rating statute. The question which I shall presently state, predicated upon the foregoing alleged facts, is fairly presented by the pleadings. This view was taken in the court below, and the question alluded to was determined on its merits. Both parties are anxious to have this important subject of controversy adjudicated by this court also, and I shall, without further discussion, assume that the matters relied on are sufficiently stated, and proceed to show why these matters do not constitute a cause of action." Southworth Case, 13 Colo., 117–18 (1889).

INDIVIDUAL PRIORITIES—"BETTER RIGHT."

(z) 86. Held by Helm, J, that the alleged facts above detailed, which were admitted by the demurrer to be true, require an answer to the following question: Does the "priority of appropriation," which, by virtue of the constitution, gives the better right, apply to individual consumers taking water through the agency of a carrier, so that, notwithstanding the pro-rating statute, each consumer acquires a separate constitutional priority of right, entitled to judicial enforcement, dating from the beginning of his specific use? If this question be answered affirmatively the statute is void, and the complaint states a cause of action; if answered in the negative the statute is in this respect valid, and the demurrer should have been sustained. Id., 119.

He answers the question in the negative in an elaborate opinion.

87. Hayt, J., does not discuss the question on its merits, but observes: "Under same circumstances, different users of water obtaining their supply through the same ditch, may have different priorities of right to the water; that appropriations do not necessarily relate to the same time. If plaintiff has alleged facts showing that he has a prior right to the use of water, which the defendants are causing to be pro-rated among those having subsequent rights, the demurrer was properly overruled, otherwise it should have been sustained;" Id. 114,

88. Helm, C. J., further says: "It is obvious from the foregoing that in my judgment all co-consumers taking water within a reasonable time have priorities of even date with each other, and the question propounded in this case revolves itself into the following: May the legislature provide that in times of scarcity, water shall be pro-rated among consumers having priorities of the same date. For if any of the co-consumers referred to in plaintiff's complaint did not use the water claimed by them within a reasonable time from the date of defendant's diversion, the fact was material and should have been pleaded. The question as thus re-stated can receive but one answer. The legislative right to provide this, as well as all other reasonable regulations, not obnoxious to constitutional objections, for the use and distribution of water cannot be denied;" Southworth case, 13 Colo., 121 (1889).

Hayt, J., says that if this were the question that it could have but one answer; Id. 116.

89. Helm, C. J., in the following language states that his views are not adopted by his associates: "I would conclude this opinion here were it not that others, including one of my colleagues on the bench, are firmly

convinced that the foregoing construction of the constitution is unsound. They contend that the constitution guarantees to each consumer a priority dating from the commencement of his individual use. The carriers' original diversion, say they, has nothing to do with the consumers priority; it is as if the consumer at the date of his use, made a distinct and independent diversion from the natural stream, merely employing for the purpose the carrier's canal; and upon this constructive diversion rests the superstructure of their theory regarding the consumer's appropriation and priority;'' Id. 124.

90. Elliott, J., says: "The question under consideration may be stated thus: Does the 'priority appropriation,' which the constitution declares 'shall give the better right as between those using the water for the same purpose,' apply to the individual consumer taking the water through the agency of an artificial stream, or is it limited to those taking water directly from the natural stream? The appropriation of water, within the meaning of the constitution, consists of two acts—first, the diversion of the water from the natural stream; and, second, the application thereof to beneficial use. These two acts may be performed by the same or different persons, but the appropriation is not complete until the two are conjoined. Hence, when the acts are performed by different persons at different times; it becomes necessary to determine which is the essential act to which the 'better right' attaches. It will be observed that by the express language of the constitution, the 'better right' is guaranteed 'as between those using the water for the same purpose.' The different purposes specified are domestic, agricultural and mechanical. Whether there are other purposes not specified need not now be discussed. Can the carrier of water for hire be said to be using the water in the sense spoken of in the constitution? The railroad company which carries farming implements from the great manufactories of the east to supply the farmers residing upon the broad prairies of the west can hardly be said to be using such implements by the mere act of thus transporting them. From the specification of the purposes for which the water may be used it would seem that the 'better right' which attaches to

the priority of appropriation was primarily intended for the benefit of those who apply the water to the cultivation of the soil or other beneficial use, rather than for the benefit of those engaged in diverting and carrying it to be used by others. The diversion and carriage of water in point of time are necessarily prior to the application of it to agricultural or other useful purposes, but they are subordinate in point of right. The former are to the latter as the means to the end, and end without which neither the diversion nor the carriage would be lawful. The carrier is the agent, the consumer is the principal. The former can lawfully pursue his occupation only by virtue of the service he renders to the latter. The consumer's right is primary and unconditional; the carrier's is secondary and dependent." Southworth case, 13 Colo., 130-1 (1889).

APPROPRIATION—NATURAL OR ARTIFICIAL STREAM.

(aa) 91. Held by Helm, C. J., that "The constitution recognizes priority only among those taking water from natural streams. The consumer himself makes no diversion from the natural stream. The act of turning water from the carrier's canal into the consumer's lateral cannot be regarded as a diversion within the meaning of the constitution; nor can this act of itself, when combined with the use, create a valid constitutional appropriation. There is, therefore, no escape from the conclusion hitherto announced by the court, that in cases like the present the carrier's diversion from the natural stream must unite with the consumer's use in order that there may be a complete appropriation within the meaning of our fundamental law." Southworth case, 13 Colo., 120 (1889).

92. Held by Elliott, J., that "A reference to sections 5 and 6, article 16, will show that it is the water of natural streams, irrespective of the mode of diversion, that is dedicated to the use of the people, subject to appropriation; and priority of right thereto is made to depend upon the time of using the water for beneficial purposes, and not upon the fact of taking the water from the natural stream. Indeed, the word "from" does not appear in either of the foregoing sections. But

it is not necessary to rely upon mere verbal analysis to
sustain the consumer's priority of right based upon
priority of use. Every consumer cannot take the water
directly from the natural stream. Irrigating ditches
and canals must be resorted to as a means of diverting
and carrying the water to places where it can be beneficially applied. No good reason can be urged why a
consumer obliged to make use of such an agency should
not be protected equally with those taking water directly
from the natural stream." Id., 131.

PRO-RATING STATUTE APPLIES TO EQUAL PRIORITIES— DITCH DECREES AS POLICE REGULATIONS.

(*bb*) 93. Held by Elliott, J., "That section 4 of the
act of 1879 (General Statutes, section 1722; same, Mills'
Annotated Statutes, 1890, section 2267), provides for
pro-rating the water actually received into and carried
by any irrigating ditch, canal or reservoir among all the
consumers therefrom in time of scarcity, so that all such
consumers shall suffer proportionately from the deficiency of water.

This provision may be properly carried into effect
when the rights of all the consumers are equal in the
matter of their respective appropriations, as when a
ditch has been constructed as a common enterprise by
and for the mutual and equal benefit of all the consumers therefrom, or when by reason of contractual
relations, waiver or other circumstances, certain consumers stand on a footing of substantial and practical
equality, having no priority of appropriation one over
another. Schilling v. Rominger, *supra*.

In Dorr v. Hammond, 7 Colo., 83, 1 Pac. Rep., 693,
it is held that a consumer may forfeit his priority of
right to the use of water by abandonment through non-
user; but where it appears as a matter of fact that one
person has been accustomed, in a lawful manner,
through the agency of an artificial stream or otherwise,
to take and apply the unappropriated waters of a natural stream to proper beneficial use, without waiver or
abandonment, from a period antedating such taking and
use by others, then to the extent of such prior taking

and use, a **prima facie** priority is established, and the statutory **regulation for** pro-rating must give **way** to the "better **right"** acquired by such priority of appropriation **under the** constitution ; and such priority must be **respected by** the ditch company, its officers and managers, **and all** others in any way connected therewith.

Giving said section 4 a literal and unqualified **interpretation, and** it manifestly conflicts with the constitution. Besides, **as** we have **seen,** the uniform decisions of this **court** plainly indicate the rule to be that, as between **those using the** water of natural streams for the **same beneficial purpose, priority** of use, gives **superiority of right irrespective of the mode of** diversion.

A single illustration will suffice to show the disastrous **consequences which would ensue if** the pro-rating **statute should be made** the **rule for** the distribution of **water for purposes of** irrigation, instead of the rule **of priority.** An irrigating ditch is constructed, the first **and only one** taking water from a small natural stream. **The first** year five consumers apply **for** and receive **each one** hundred inches of water for the irrigation of **their** lands; **the** next year, the ditch being enlarged, five more apply and receive a like quantity, and the third year five more, and so on successively, until thirty **or** forty consumers are located under the ditch. Perhaps the first **five** might **be** required to pro-rate with each other in time of **scarcity should** their appropriations be practically equal in **point of time ;** but under **the statute** the first five **would also be** compelled **to pro-rate with all subsequent consumers,** until the **amount of water that each would receive would** be **so infinitessimally** small as to **be of no practical** value, **and would be** eventually entirely **wasted before** it could **be applied."** Southworth **case, 13 Colo., 135,** etc.

94. Held by Helm, **C. J., that** the **said** pro-rating statute "reaches all consumers having secured priorities through diversion by carriers **alike. It** makes no distinction among them. Each **and** all are equally within **its purview.** This is purely a question of constitutional construction, and the constitutional meaning **does not seem to** be obscured by **any serious** ambiguity ; **but were**

the meaning doubtful, the argument based upon supposed
hardship and injustice, is in my judgment, not entitled
to notice.   It is true the consumer, who first uses, may
be compelled to pro-rate with another whose use is subsequent in date, but each consumer has a perfect right
to go to the natural stream for the water he needs.
There is no law forcing him to deal with the carrier   It
is no answer to say that the over-powering law of necessity takes away his volition to choose, for he in fact
makes his election when he purchases land so far from
the natural stream as to compel reliance upon the carrier.   But when he elects to take from the carrier's
canal, and thus to employ this lawful agency he cannot
reject the accompanying lawful obligations.   The legislature is powerless to say that he shall not take unappropriated water from the natural stream; but that body
can declare that if he employs the services of a carrier
he shall take notice of and be governed by such valid
regulations as have been adopted, pertaining to the distribution of water therefrom.   Under the constitution,
statutes and decisions as I read them, the consumer
takes with full knowledge that the carrier's entire diversion will ripen into valid appropriations, provided the
water be applied within a reasonable time to beneficial
uses.   He also takes with knowledge that the different
lawful co-consumers will have the same priority, a priority resting for its commencement upon the carrier's
diversion, or dating from a subsequent enlargement of
the quantity of water to which the carrier was originally
entitled.   He must, therefore, be presumed to know that
in times of scarcity his use may be subjected to two
interruptions, viz.: first, that canals and ditches holding
priorities antedating the diversion of his carrier may
demand all the water in the natural stream, so that there
will be none for him or any of his co-consumers ; and
second, that if there is water, but not the full quantity
appropriated, he will be obliged to pro-rate with such
co-consumers.   Under these circumstances the consumer
is hardly in position to resist the enforcement of the pro-
rating statute, or to assert that it operates harshly and
unjustly upon him."   Southworth case, 13 Colo., 122-3.

95. Held by Elliott, J., "That the objects and purposes of the acts of 1879 and 1881, providing for the
adjudication of priorities, (L. '79, p. 94, etc.; L. '81, p.
142, etc.; same Mills' Ann. Stat. 1890, secs. 2399-2439,)
is, by way of police regulations, to secure the orderly
distribution of water for irrigation purposes, and to provide a system of procedure for determining the priority
of rights as between the carriers, or strictly speaking,
priority of diversion as between themselves. They are
using the water " for the same purpose and by analogy
may be termed 'appropriators' or 'quasi appropriators.'
These acts are to protect their rights and to prevent the
inevitable conflicts that would occur if diversion from
the natural stream were not under government control."
Id., 134, etc.

96. Held by Helm, J., that the foregoing theory as
to the purpose of the acts of 1879 and 1881, reads well,
but its practical application is not feasible, and its
results in making all the said acts worse than useless if
not void. Id., 127.

Section 512. Right of way for ditches, flumes.
Sec. 7. All persons and corporations shall have the
right of way across public, private and corporate lands
for the construction of ditches, canals and flumes, for
the purpose of conveying water for domestic purposes,
for the irrigation of agricultural lands, and for mining
and manufacturing purposes, and for drainage, upon
payment of just compensation.

Section 513. County commissioners fix rates for
water. Sec. 8. The general assembly shall provide by
law that the board of county commissioners in their respective counties, shall have power, when application
is made to them by either party interested, to establish
reasonable maximum rates to be charged for the use of
water, whether furnished by individuals or corporations.

COLORADO DECISIONS AND CITATIONS.

2. This section referred to; Knoth v. Barclay, 8
Colo., 303 (1885).

3. Agriculture in Colorado depends on irrigation; Yunker v. Nichols, 1 Colo., 553 (1872).

(a) STATEMENT OF YUNKER–NICHOLS CASE—STATUTE OF FRAUDS.

4. This was an action of trespass brought by Yunker v. Nichols, because the latter living higher up on a ditch built by their joint labor, appropriated all the water in such ditch to his own use while the same was flowing over his own land, and left no water in the ditch to flow on to plaintiff's lands, by means whereof the plaintiff's growing crop was greatly injured and diminished in value. The case was reversed by the supreme court, because the jury was instructed that if Yunker's right to have the water flow on to his land through a ditch over the land of Nichols was conferred only by verbal agreement and never reduced to writing, as required by the statute of frauds, the plaintiff could not recover; Id. 552.

5. In England and in this country it is considered that the right of one person to conduct water over the land of another is an interest in real estate which must be conveyed by deed in compliance with the statute of frauds. Hallett, J., in Id., 552.

6. But in Colorado and the arid region all lands are held in subordination to the dominant right of others, who must necessarily pass over them to obtain a supply of water to irrigate their own lands, and this servitude arises not from grant, but by operation of law. A deed to a right of way for such irrigating ditch is not necessary Id., 555.

7. Under these principles consent to build a ditch gives a right of way, and if consent is withheld a party can proceed under the statute. Id.

EXECUTED LICENSE—REVOCABLE OR IRREVOCABLE.

(b) 8. The states of Ohio and Pennsylvania (Wilson v. Chalfant, 15 Ohio, 253; Wents v. De Haven, 1 Serg. & R., 312) adhered strictly to the doctrine of the irrevocability of an executed license, while in all other states, as well as in England (Wood v. Leadbitter, 13 M. & W., 838), such licenses have been held revocable at will, being in contravention of the statute of frauds; but the Ohio and Pennsylvania rule is followed there as the more equitable, thus preventing the statute of frauds from being fraudulently and oppressively used, and this was also formerly the rule in Maine, New Hampshire, Iowa and Indiana. Belford, J., in Id., 558–565, Ricker v. Kelly, 1 Greenl., 117; Pitmandy Poor, 38 Me., 237; Woodbury v. Parshley, 7 N. H., 237; Ameriscoggin Bridge v. Bragg, 11 Id., 108; Sampson v. Burnside, 13 Id., 264; Houston v. Laffer, 46 Id., 507; Wickersham v. Orr, 9 Iowa, 260; Beatty v. Gregory, 17 Id., 109; Snowden v. Wilds, 19 Ind., 14.

ESTOPPEL AS APPLIED TO AN EXECUTED LICENSE TO BUILD A DITCH.

(c) 9. The irrevocability of such an executed license in the building of an irrigating ditch is also deducted from the doctrine of *estoppels in pais*. He who by his conduct or admissions induces another to act cannot afterward be permitted to assert the contrary to the injury or prejudice of the party who has already acted upon the faith and in the belief created by him. Belford, J., in Yunker v. Nichols, 1 Colo., 652 (1872); Wells, J., dissents as to this point; Id., 569; *contra*, Stewart v. Stephens, 10 Id., 446–8 (1887); compare Burnham v. Freeman, 11 Id., 606 (1888).

10. The above principal of *estoppel in pais* applies to the case of an irrigating ditch, which is an interest or easement in land, the same as the private way over the land of another. Belford, J., in Yunker v. Nichols, 1 Colo., 562–568 (1872). This principle extends to real estate and affects the title thereof as well as to personality Id.; Hern v. Rogers, 9 B. & C., 577; Farr v. Newman, 4 D. & E., 636; Smith v. Doe, 7 Price, 509; Shaw v. Bebee, 35 Vt., 208; Wents v. De Haven, 1 Serg. & R., 312; Corbet v. Norcross, 35 N H., 115; Heard v. Hall, 16 Pick., 457; White v. Perkins, 24 Id., 324; Sharan v. Mennick, 6 Nev., 389; Kelly v. Taylor, 23 Cal., 11.

11. Courts would betray their trust if, in the administration of law, or the expounding of the constitutional principles, they shut their eyes and refuse to recognize those conditions of society which call into force and operation principles whose existence and recognition cannot be disregarded without bringing ruin on all; Belford, J., in Yunkers v. Nichols, 1 Colo., 569 (1872).

DITCH—RIGHT OF WAY, NOT FROM STATUTE BUT FROM NECESSITY.

(*d*) 12. The right of every proprietor to have a way over the lands intervening between his possessions and the neighboring stream for the passage of water for irrigation of so much of his land as may be actually cultivated, is well sustained by force of the necessity arising from local peculiarities of climate. This right springs out of the necessity, and existed before the statute (L. '61, p. 67; same R. S., 68, p. 363; same Mills' Ann. Stat., 1890, sec. 2257) was enacted and would still survive though the statute was repealed. Dissent is entered to the deduction of such right from the statute; Wells, J., in Yunkers v. Nichols, 1 Colo., 570 (1872).

13. Thatcher, C. J., says that the right to convey water over the lands of another, for irrigating purposes, is founded on the imperious law of nature, to which, it must be presumed, the government parts with its title; Schilling v. Rominger, 4 Colo., 109 (1878).

14. This servitude over the lands of another may, therefore, be created otherwise than by deed: Id.

THERE MUST BE COMPENSATION FOR RIGHT OF WAY.

(*e*) 15. The constitution was adopted since the above decision of Yunker v. Nichols was made, and this section and art. 2, secs. 14 and 15 prohibit the taking of private property for private use without compensation; and as shown in Tripp v. Overocker, 7 Colo., 73, the legislature has provided the proceedings by which private property may be subjected to private use; Stewart v. Stephens, 10 Id., 445-6 (1887).

16. Under this **section** and **also Mills'** Ann. Stat.,
1890, **sec.** 2257-8, a party **is** given **a right of** way for a
ditch over the lands of another **upon the** payment of
just compensation therefor; Tripp v. Overocker, 7 Colo.,
73 **(188**3); Burnham v. Freeman, 11 Id., 607 (1888).

17. Sec. 1716, G. S., 1883 (same Mills' **Ann. Stat.,**
1890, **sec.** 2261) is not in conflict with **this section of**
the constitution, for it recognizes the **right of way for**
ditches **and** seeks only to regulate the **exercise of such**
right so **as to** inflict the least possible injury **and incon-**
venience upon the **owner** of the servient estate; **Tripp**
v. Overocker, 7 Colo., 73 (1883).

18. **The right of a purely private party to con-**
demn a right of **way** for a **ditch to convey water to his**
lands for domestic, agricultural **and mining purposes is**
guaranteed by this section, **and section** 14, **article 2;**
the statute regulates the manner of **exercising the right.**
Id. Downing v. More, 12, Colo., 318 **(1888).**

19. The latter case modifies **the former, but not**
as to this or any constitutional point.

20. As to the right of **way for** ditch, **see** Knoth v.
Barclay, 8 Colo., 303 (1885), in **note to** article **2, section**
15.

RIGHT TO CONDEMN USE OF DITCH **CONSTRUCTED.**

(*f*) **21.** A party having the right **to the** use of
water, **can** condemn a right of way for his ditch, or condemn **the** right to use the one already constructed.
Burnham v. Freeman, 11 Colo., 607 (1888).

22. An interest **in an irrigating ditch is realty and**
cannot pass by a verbal **sale. Id.,** 606 **(1888);** Smith v.
O'Hara, 43 Cal., **371.**

Section 513. County commissioners **fix** rates **for**
water. Sec. 8. The general assembly **shall** provide by
law that the board of **county** commissioners, in their
respective counties, shall have power, when application
is made to them by either party interested to establish
reasonable maximum rates to be charged for **the use of**
water, **whether** furnished **by** individuals **or corpora-**
tions.

1. Water rates fixed by county commissioners, see Mills' Ann. Stat., 1890, sections 570, 2295, 2298.

I. COLORADO DECISIONS AND CITATIONS.

2. Farmers are generally too poor to build main ditches of their own, hence individuals and corporations engage in the business of building and operating these mains and furnishing water to the farmers along the lines thereof. If these persons and corporations were entirely uncontrolled in the matter of prices, injustice and trouble would follow. It is exceedingly proper they should be subjected to reasonable regulations, as provided in this section of the constitution. Golden Canal Co. v. Bright, 8 Colo., 148 (1884).

3. The price to be charged is involved in the regulations of the use of water. Id., 149.

4. While it is true there is opportunity for gross injustice to the ditch owner or the consumer, if an appeal is not allowed from the decision of the county commissioners, fixing maximum rates, still, as a matter of fact, the act (L. '79, p. 94, etc.) does not provide for such appeal, nor does any other act. Id., 155.

CANAL COMPANY, RIGHT TO CARRY FOR HIRE RECOGNIZED.

(a) 5. This section unquestionably contemplates and sanctions the business of transporting water for hire from natural streams to distant consumers. Wheeler v North Colo. Irr. Co., 10 Colo., 588 (1887)

6. The board of county commissioners are a judicial or quasi-judicial tribunal when fixing the rates under this section. Id., 589.

7. By the common law, even if the constitution and statutes were absolutely silent on the question of charges for transportation, and the times and manner of their collection, the same must be reasonable, and they are subject to reasonable judicial control. Id., 589-90.

8. The enforcement on the part of a canal company of unreasonable and oppressive demands, in rela-

tion to the time and manner of collecting rates, is by fair implication forbidden by this section. Id., 590.

9. For the fixing of maximum rates would be grossly inadequate protection, if either party might dictate absolutely the time and condition of payment. Id.

10. The primary objects of this section were to encourage and protect the beneficial use of water; and, while recognizing the carrier's right to a reasonable compensation for its carriage, collectable in a reasonable manner, the constitution also unequivocally asserts the consumer's right to its use, upon payment of such compensation. Id.

11. Any unreasonable regulations or demands that operate to withhold or prevent the exercise of this constitutional right by the consumer must be held illegal, even though there be no express legislative declaration on the subject. Id.

12. When the canal company has fixed a rate of its own, with which the consumer is satisfied, it is not necessary to apply to the county commissioners to fix the rate; and G. S., '83, sec. 311, (same Mills' Ann. Stat., 1890, sec. 570,) must be so understood. Id., 595.

13. Against illegal exactions the consumer's only remedy is not by application to the county commissioners to fix the rates. They can fix the maximum amount, but not the time or conditions of payment; the legislature may fix the latter. Id.

Quotations from constitution end here.

STATUTES.

Section 570. When compelled to furnish water. Any company constructing a ditch under the provisions of this act shall furnish water to the class of persons using the water in the way named in the certificate in the way the water is designated to be used, whether miners, mill men, farmers or for domestic use, whenever they shall have water in their ditch unsold, and shall at

all times give the preference to use of the water in said
ditch to the class named in the certificate, the rates at
which water shall be furnished to be fixed by the
county commissioners as soon as such ditch shall be
completed and prepared to furnish water. G. L. '77, p.
172, sec. 277; G. S. '83, p. 199, sec. 311.

1. County commissioners fix rates of water. See
Colo. Const., art. XVI., sec. 6.

2. Rates of charge for water and regulation
thereof. Sec. 2295, etc.

3. Right to continue purchasing water. Sec.
2997.

4. This section is not repealed by sec. 2297.
Wheeler v. North Colo. Ir. Co., 10 Colo., 595 (1887).

5. This section expressly commands ditch companies having water in their canals not taken to furnish
the same to the class of persons using it in the manner
named by the articles of incorporation. The declaration therein that this rate shall be fixed by the county
commissioners must be taken with the constitutional
condition attached. Id.

Section 571. Shall keep ditch in repair. Every
ditch company organized under the provisions of this
act shall be required to keep their ditch in good condition, so that the water shall not be allowed to escape
from the same, to the injury of any mining claim, road,
ditch or other property; and whenever it is necessary to
convey any ditch over, across or above any lode or mining claim, or to keep the water so conveyed therefrom,
the company shall, if necessary to keep the water of
said ditch out, or from any claim, flume the ditch so far
as necessary to protect such claim or property from the
water of said ditch. G. L. '77, pp. 172, 173, sec. 278;
G. S. '83, p. 199, sec. 312.

1. Owner of ditch to maintain embankment.
Sec. 2274.

2. Vested rights of mill and ditch owners.
Sec. 2275.

3. Liability of owner for damages. Sec. 2272.

4. A ditch company is liable for damages caused by allowing water to overflow the banks of their ditch and flood the land of another. The liability arises from the failing to exercise ordinary care in preventing the escape of the water. Greeley Irr. Co. v. House, 24 Pac. Rep. 330 (1890); 14 Colo., —; Ditch Co. v. Anderson, 8 Id., 131; Water Co. v. Middaugh, 12 Id., 443.

Section 572. Consolidation of ditch companies. Companies organized under the laws of this state holding ditches or canals by virtue of their organization, which derive their supply of water for their respective ditches or canals from the same headgate or gates, or the same source or sources of supply, may consolidate their interests and unite their respective companies under one name and management, by filing a certificate of that fact in the office of the secretary of this state, and a counterpart thereof in the office of the recorder of the county or counties in which such ditch or canals are situated; which certificate shall be signed by the presidents of the companies so uniting, with the common seals of the companies affixed thereto; and shall set forth the fact of such union of interests, and give the name of the new company thus formed. L. '76, pp. 68, 69, sec. 1; Omitted, G. L., '77; G. S., '83, pp. 199, 200, sec. 313.

Section 573. Shall commence work within ninety days—Complete in two years—Forfeit—Ditch three years. Any company formed under the provisions of this act for the purpose of constructing any ditch, flume, bridge, ferry or telegraph line shall within ninety days from the date of their certificate, commence work on such ditch, flume, bridge, ferry or telegraph line, as shall be named in the certificate, and shall prosecute the work with due diligence until the same is completed, and the time of the completion of any such ditch, bridge, ferry or telegraph line shall not be extended beyond a period of two years from the time work was commenced as aforesaid; and any company failing to commence work within ninety days from the date of the certificate, or failing to complete the same within two years from the time of commencement as aforesaid, shall forfeit all rights to the water so claimed, and the same shall be subject to be claimed by any other company; the time for the completion of any flume constructed under the provisions of this act shall not be extended beyond a period of four years; *provided*, This section shall not apply to any ditch or flume for mining or other purposes, constructed through and upon any ground owned by the corporation; and provided further, that any company formed under the provisions of this act to construct a ditch for domestic, agricultural, irrigating, milling and manufacturing purposes, or any part or either thereof, shall have three years from the time of commencing work thereon within which to complete the same, but no longer. G. L., '77, pp. 179, 180, sec. 296; G. S., '83, p. 200, sec. 314.

Section 574. Damaging road, ditch, flume—Penalty. Any person who shall willfully or maliciously damage or interfere with any road, ditch, flume, bridge, ferry or telegraph line, or any of the fixtures, tools, implements, appurtenances, or any property of any company which may be organized under the provisions of this act, upon conviction thereof before any court of competent jurisdiction in the county where the offense shall have been committed, shall be deemed guilty of a misdemeanor, and shall be punished by fine or imprisonment or both, at the discretion of the court, said imprisonment not to exceed one year and said fine not to exceed five hundred dollars, which fine shall be paid into the county treasury for the use of the common schools, and said offender shall also pay all damages that any such corporation may sustain, together with costs of suit. G. L. '77, p. 180, sec. 297 ; G. S. '83, p. 200, sec. 315.

1. Penalty for cutting or breaking gate, bank, side of ditch, flume, etc., sec. 2393.

Section 575. What certificate shall specify. When any company shall organize, under the provisions of this act, to form a company for the purpose of constructing a flume, their certificate, in addition to the matters required in the second section of this act, shall specify as follows: The place of beginning, the terminus, and the route so near as may be, and the purpose for which such flume is extended, and when organized, according to the provisions of this act, said company shall have the right of way over the line proposed in such certificate for such flume; *Provided*, It does not conflict with the rights of any former fluming, ditching, or other company. G. L. '77, p. 173, sec. 279; G. S. '83, pp. 200, 201, sec. 316.

1. Duties of owners, sec. 2274, etc.

Section 576. What certificate shall specify—Place —Stream—Ownership. When three or more persons shall associate, under the provisions of this act, to form a company for the purpose of constructing a bridge, or establishing a ferry over any of the streams of water in this state, their certificate, in addition to the matters required in the second section of this act, shall specify as follows: The place where such bridge, or places at which such bridges or ferry is to be built or established, and on what streams, and that the banks on both sides of the stream where the said bridge or ferry is to be built or established are owned by said company, or that they have obtained in writing the consent of the owners of the banks where the said bridge is to be built, to erect said bridge, or establish the said ferry as aforesaid, or that the banks at such place are a public highway. G. L., '77, p. 173, sec. 280; G. S., '83, p. 201, sec. 317.

1. Fords and ferries, secs. 579, 580.

DIVISION V.

WATER PRIVILEGES.

Section 949. Commissioners may subscribe to capital stock of corporations and issue bonds. It shall be lawful for the board of commissioners of any county in this state to subscribe to the capital stock of any incorporated company organized under the laws of this state for the purpose of constructing ditches, flumes, or other works for the supply of such county with water for mining, milling, irrigating and domestic and fire purposes, such subscription to be paid by the issue of the bonds of said county, as herein provided. L. '74, p. 193, sec. 1; G. L. '77, p. 638, sec. 1839. This act is omitted from G. S. '83.

Section 950. How such aid from counties may be obtained—Special election. Whenever any such incorporated company shall solicit the aid of such county by subscription to its capital stock, it shall submit to the board of county commissioners a statement in writing, to be filed in the office of the county clerk of such county, setting forth the sources from which water is to be obtained, and the proposed capacity of the the ditch, flume or pipes by which the water is to be brought, together with the number and size of the reservoirs to be constructed, and the route, as near as practicable, over which the same is to be brought, the estimated cost of the said works when completed, and the rates at which they agree to furnish water for the purposes set forth for the first three years after the same is in operation. Such statement shall also be accompanied with a petition of at least fifty legal voters of said county, who shall have paid taxes on property, real or personal, in said county during the year preceding that in which such petition is drawn, requesting said board to call an election in said county upon the question of the issue of the bonds of such county in aid of said company, in payment of the proposed subscription to the capital stock of said company. Upon the receipt of such statement and petition, the said board of county commissioners shall thereupon, within ten days thereafter, call a special election upon such question, and enter an order on their journal thereof. Such election shall be upon notice published in some newspaper published in said county, for at least three weeks before the day named therein upon which the vote shall be taken; or, if there is no newspaper published in said county, then by putting notices at the several places of voting in the different precincts of said county for the same period. Said notice shall contain the statement of the said company as above prescribed, together with the terms and conditions upon which such stock is to be subscribed, and the bonds issued, and any other matter necessary to a fair, impartial and intelligent expression of the will of the voters of such county upon the question submitted; which question shall be as to the subscription to the capital stock of said company and the issuance of the bonds of said county in payment thereof. Such notice shall also state the time which such bonds shall run, the rate of interest they shall bear, and the manner in which they shall be paid. L. '74, pp. 193-195, sec. 2 ; G. L. '77, pp. 638-639, sec. 1840.

Section 951. Manner of issuing bonds. If two-thirds majority of all votes cast at such election shall be in favor of the subscription to the said stock and the issuance of such bonds, it shall be the duty of the said board of county commissioners to subscribe said stock and issue said bonds of said county, and to exchange the same at the par value for the stock of said company at its par value; *provided*, That no bonds shall be issued bearing interest at a rate exceeding ten per cent. per annum; and provided further, that no bonds shall be issued due and payable until fifteen years after the date thereof, except at the option of the said board of county commissioners of such county, after five years from date thereof. L. '74, p. 195, sec. 3 ; G. L. '77, p. 639, sec. 1841.

1. This section referred to in Coulter v. Routt County, 9 Colo., 264 (1886).

Section 952. Special tax for payment of principal and interest. The board of county commissioners of any such county shall have power to levy a special tax, to be paid in cash, of not to exceed three mills on each and every dollar of property assessed and liable to taxation in such county, each year, for the payment of the interest annually on such bonds; and at the end of five years it shall be the duty of such board to levy a tax not exceeding five mills on the assessed value of the property in said county, for each and every year, for the payment of the interest and principal of said bonds; and all money applicable to the payment of the principal of said bonds shall be applied to the payment thereof by the said board at the end of the fiscal year of each county; *provided*, That the amount so levied each year shall be an amount sufficient at such rates to pay the amount of said bonds at maturity. L. '74, p. 195, sec. 4; G. L. '77, pp. 639, 640, sec. 1842.

Section 953. Discrimination in water rates. No incorporated company to whose stock any county has subscribed shall make any discriminating rates against or in favor of any person or corporation, or charge one person or corporation more for a given amount of water for a given purpose than another, except that in cases where a small quantity of water only is required it shall be lawful for the company to make such charges as may be just and reasonable, without regard to the rates fixed for other purposes and in larger amounts. L. '74; pp. 195, 196, sec. 5; G. L. '77, p. 640, sec. 1843.

Section 954. Exchange of stock for bonds. It shall be lawful, at any time after three years from the date of any bonds issued under the provision of this act, for the board of county commissioners of any county to exchange the stock so held and subscribed by said county for the bonds of said county, such stock to be exchanged at its par value and such bonds taken at their par value. L. '74, p. 196, sec. 6; G. L. '77, p. 640, sec. 1844.

Section 955. Limitation of amount of bonds issued. The amount of bonds issued by any county under the provisions of this act shall in no case exceed four per cent. of the assessed value of the property situate in said county for the year preceding that in which such bonds are voted. L. '74, p. 196, sec. 7; G. L. '77, p. 640, sec. 1845.

Section 996. Arapahoe county not included. The provisions of this act shall not apply to Arapahoe county. L. '74, p. 196, sec. 8; G. L. '77, p. 640, sec. 1846.

Section 1301. Taking illegal fees—Triple damages—Penalty. Any judge, justices of the peace, clerk, sheriff, constable, city marshal, or other public officer, who for the performance of an official duty, for which a fee or compensation is allowed or provided by law, shall willfully and knowingly demand or receive any greater fee or compensation either in money or other thing of value than what is allowed or provided by law for the same, or who shall willfully and knowingly demand or receive any such fee or compensation where no fee or compensation whatever is authorized or prescribed by law, shall be guilty of a misdemeanor, and upon conviction thereof shall be confined in jail not less than one nor more than six months, and shall be fined not less than one hundred, nor more than five hundred dollars, besides being liable on a civil action to the person or persons from whom such fee or compensation is thus knowingly and illegally demanded or received, for three times the value or amount thereof, and upon the examination or trial of such offense, the defendant shall be presumed to have acted willfully and knowingly, until the contrary is shown. L. '74, p. 166, sec. 1 ; G. L. '77, p. 426, sec. 1159; G. S. '83, p. 324, sec. 817.

Section 1376. Polluting streams—Penalty. If any person or persons shall hereafter throw or discharge into any running stream of water, or into any ditch or flume in this state, any obnoxious substance, such as refuse matter from slaughter house or privy, or slops from eating houses or saloons, or any other fleshy or vegetable matter which is subject to decay in water, such person or persons shall upon conviction thereof, be punished by a fine not less than one hundred dollars nor more than five hundred dollars for each and every offense so committed. L. '74, p. 99, sec. 1; G. L. '77, p. 307, sec. 760; G. S. '83, p. 342, sec. 882.

Section 1716. Petition—Parties—When state is party—Private property taken when. That in all cases where the right to take private property for public or private use without the owner's consent, or the right to construct or maintain any railroad, public road, toll road, ditch, bridge, ferry, telegraph, flume, or other public or private work or improvement, or which may damage property not actually taken, has been heretofore, or shall hereafter be conferred by general laws or special charter, upon any corporate or municipal authority, public body, officer or agent, person or persons, commissioner or corporation, and the compensation to be paid for in respect of the property sought to be appropriated or damaged for the purposes above mentioned, can not be agreed upon by the parties interested; or in case the owner of the property is incapable of consenting, or his name or residence is unknown, or he is a non-resident of the state, it shall be lawful for the party authorized to take or damage the property so required, or to construct, operate and maintain any railroad, public road, toll road, ditch, bridge, ferry, telegraph, flume, or other public or private work or improvement, to apply to the judge of the district or county court, either in term or vacation, where the said property or any part thereof is situate, by filing with the clerk a petition, setting forth by reference:

1.   His or their authority in the premises.

2.   The purpose for which said property is sought to be taken or damaged.

3.   A description of the property.

4.   The names of all persons interested therein as owners or otherwise as appearing of record, if known, or if not known, stating the fact.

5.   And praying such judge to cause the compensation to be paid to the owner to be assessed.

6.   If the proceedings seek to effect the property of persons under guardianship, the guardians or conservators of persons having conservators, shall be made parties defendant, and if of married women, their husbands shall also be made parties.

7.   Persons interested, whose names are unknown, may be made parties defendant by the description of the unknown owners.

8. But in all such cases an affidavit shall be filed by or on behalf of the petitioner, setting forth that the names of such persons are unknown.

9. In cases where the property is sought to be taken or damaged by the state for the purpose of establishing, operating or maintaining any state house, or charitable or other state institution of improvement, the petition shall be signed by the governor, or such other person as he shall direct, or as shall be provided by law. Under the provisions of this act, private property may be taken for private use, for private ways of necessity, for reservoirs, drains, flumes or ditches, on or across the lands of others for agricultural, mining, milling, domestic or sanitary purposes.

The amendment of said act shall not be construed to affect any right, either as to remedy or otherwise, nor to abate any suit or action' or proceeding existing, instituted or pending under the act so hereby amended. L. '85, pp. 200, 201, sec. 1, amending G. L., '77, pp. 397, 398, sec. 1059; G. S., Code, '83, p. 75, sec. 238.

DIVISION I.

RIGHT OF WAY—APPROPRIATION—USE OF WATER.

Section 2256. Owners of land on streams entitled to use water—Appropriation. All persrns who claim, own or hold a possessory right or title to any land or parcel of land within the boundary of the state of Colorado, as defined in the constitution of said state, when those claims are on the bank, margin or neighborhood of any stream of water, creek or river, shall be entitled to the use of the water of said stream, creek or river for the purposes of irrigation, and making said claims available to the full extent of the soil, for agricultural purposes. L. '61, p. 67, sec. 1; R. S., '68, p. 363, sec. 1; G. L., '77, p. 515, sec. 1372; G. S., '83, pp. 560, 561, sec. 1711.

1. As to wasting water, see secs. 2282-3.

2. Statement to be filed with county clerk by owner, when capacity exceeds one cubic foot, sec 2265.

3. Damage done to ditches, flumes, etc., by floating timber. See sec. 2013.

4. For irrigation of towns and cities. See sec. 4539, etc.

5. As to irrigation for counties. See Div. V., chap. 33, "County Government," secs. 949-956.

6. As to the measure of water. Statute inch—see sec. 4643; cubic foot—sec. 2467

7. The chapter on "Drainage" (G. S., '83, p. 399, etc.) was repealed by the L., '85, p. 190, sec. 1. As to the rights of an appropriator to enter land of another to fix ditch, etc., see sec. 2264, note.

8. Irrigation defined. See const., art. XVI., sec. 5, note 22, etc.

9. As to fixing maximum rates for transporting water. See sec. 2295, etc.

10. As to the adjudicating of rights. See sec. 2399, etc.

11. As to seepage, percolating and underground waters. See sec. 2269 and notes.

12. This section is valid and constitutional. Yunker v. Nichols, 1 Colo., 566 (1872).

13. For the points in this case. See const., art. XVI, sec. 7, notes 4-14.

14. This section commented on in Coffin v. Left-Hand Ditch Co., 6 Colo., 450-1 (1882). See also sec. 568 and notes.

15. See as to the construction of the words in this section, "on the bank, margin or neighborhood." Sec. 568, note 9.

ABANDONMENT.

See const., art. XVI, sec. 6, note 39, etc.

23. An appropriator of water, who for many years makes no use of the water, allows his ditch to become obliterated, and interposes no objection to the diversion of the water by a subsequent appropriator, will be presumed to have abandoned his right of priority. Dorr v. Hammond, 7. Colo., 79; Farmers' H. L. Canal & R. Co. v. Southworth, 13 Id., 136 (1889).

25. A failure to use for a time is competent evidence on the question of abandonment, and if such non-use be continued for an unreasonable period it may fairly create a presumption of intention to abandon; but this presumption is not conclusive, and may be overcome by other satisfactory proofs. Id.; Sieber v. Frink, 7 Colo., 148.

APPROPRIATION—PRIORITY GIVES BETTER RIGHT.

27. For all the Colorado cases upon this subject see Const., art. XVI., sec. 6, and notes. For the constitutional provisions of the several western states on irrigation see Const., art. XVI., sec. 5, notes.

28. The first appropriator of water from a natural stream for a beneficial purpose has a prior right thereto to the extent of such appropriation. Wheeler v. North Colo Irr. Co., 10 Colo., 582; S. C., 3 Am. St. Rep., 605; Hammon v. Rose, 11 Colo., 524–5 (1888); S. C., 7 Am. St. Rep. 258.

28 (a) If land be rendered productive by the natural overflow of water thereon, without the aid of any appliances whatever, the cultivation of such land by means of the water so naturally moistening the same is sufficient appropriation of such water to the amount necessary for such use. Thomas v. Guiraud, 6 Colo., 532 (1883).

29. The diversion of the water of a stream with the object of draining simply, or without the intention of applying them to some useful purpose, does not constitute an appropriation. Thomas v. Guiraud, 6 Colo., 530.

31.   To constitute a legal appropriation the water claimed must be applied to some beneficial use or purpose.   See Const., art XVI., sec. 6, notes 1–15.

34.   The legislature cannot prohibit the appropriation or diversion of unappropriated water for useful purposes from natural streams upon the public domain, but it may regulate the manner of affecting such appropriation or diversion, and may designate how the water shall be turned from the stream or how it shall be stored and preserved.   Larimer County R. Co. v. People, 8 Colo., 614.

APPROPRIATION—DILIGENCE.

44.   To acquire a right to water from the date of the diversion thereof, one must within a reasonable time employ the same in the business for which the appropriation is made.   What shall constitute such reasonable time is a question of fact, depending upon the circumstances connected with each particular case.   See Const., art. XVI, sec. 6, notes 24, etc., Highland Ditch Co. v. Mumford, 5 Colo., 336, 1880.

APPROPRIATION COMPLETED—RELATES BACK.

48.   Although the appropriation is not deemed complete until the actual diversion or use of the water, still if such work be prosecuted with reasonable diligence, the right relates to the time when the first step was taken to secure it.   See Const., art. XVI., sec. 6, note 24, etc.

APPROPRIATION—PLACE OF USE.

50.   The right to water acquired by prior appropriation is not in any way dependent upon the locus of its application to the beneficial use designed or to the particular use to which it was first applied.   See Const., art. XVI., sec. 6, notes 27-38; Coffin v. Left-Hand Ditch Co., 6 Colo., 443; Thomas v. Guiraud, Id., 530.

APPROPRIATION—POINT OF DIVERSION.

51.   One entitled to divert a quantity of water from a stream may take the same at any point on the stream

and may change the point **of diversion if** the rights of
others be not injuriously **affected** by the change.   See
Const., art. XVI., **sec. 6, notes 27-38;** Sieber v. Frink, 7
Colo., 148.

APPROPRIATION—EXTENT OF.

54. The appropriation of the water of a **stream for**
a particular purpose is an appropriation of only **so much**
of the water as is necessary for that purpose, and **the**
surplus, **if any, may** be taken by others.   Sieber **v.**
Frink, 7 **Colo., 148,**

59. A person **can be subrogated to the rights of an**
original appropriatior or **his grantee, to a certain num-**
ber of inches of **water.   Such right passes by grant.**
Bugh v. Rominger, **24 Pac. Rep., 1046** (1890); **14 Colo.,**
see Const., art. XVI., sec. 6, **notes 16, etc.**

60. **The right acquired by the prior appropriator
is limited to the** amount **of water appropriated.   In sub-**
ordination to his right **thus limited, others may appro-**
priate the remainder of the water **running in the stream.**
Thomas v. Guiraud, 6 Colo., **530.**

62. A subsequent appropriator from **a natural**
stream has no right to destroy the ditch of a **prior** appropriator, or to materially diminish the quantity or
deteriorate the quality of the water to which the latter
is entitled ; nor has the prior appropriator a right **to**
extend his use of **water** to the prejudice of the subsequent appropriator.   **Sieber v.** Frink, 7 Colo., 148; Larimer Co. R. C. v. People, **ex rel.,** 8 Id., 614; see Const.,
art. XIV., sec. 6, notes, **31, etc.**

APPROPRIATION AND COMMON LAW.

64. **See Const., article XVI., section 6, notes 5**
and 19-20.

APPROPRIATION BY DAM OR RESERVOIR.

69. **One** may utilize **as a** reservoir for storing
water a natural depression **on the** public land, which
includes the bed of a stream; but he must see to it that
no legal right of prior appropriators **or other** persons is

in any way interfered with by his acts.   Larimer Co. R.
Co. v. People ex rel., 8 Colo., 614.

CONVEYANCES—DEEDS—APPURTENANCES—WATER RIGHTS.

76. It is acquired by use and **not grant.   See**
Mills' Const. Anno., **art.** XVI., sec. 6, **notes** 16, etc.

DITCHES.

94. A municipal corporation which accepts **the**
dedication of streets across which a ditch has been **pre-**
viously located and right of way therefore **acquired,**
takes the same **subject** to the prior rights of the owners
of **the** ditch; and **when** the necessities of the public require that such **ditch be** bridged at the street crossings,
it is the duty of the city, and not of the owner of the
ditch, to construct such bridges.   Denver v. Mullen, 7
Colo., 345.

98. A water ditch and the water-right appurtenant
thereto are real property; see Mills' Const. Ann., art.
XVI., sec. 8, note 22.

109. As to irrigation and ditch companies, see sec.
567, etc.; flume companies, see sec, 575.   They are
common carriers or impressed with a public trust.
Const., art. XVI., sec. 6, notes 62, etc.

110. Ditch owners, as such, are carriers and must
furnish water at the established rate (the county commissioners being empowered to fix a maximum rate) to
the class of persons using it in the manner named in
the articles of incorporation.   Golden Canal Co. v.
Bright, **8 Colo.,** 144; Wheeler v. Northern Colo. Irr. Co.,
11 Id., **582;   S. C.,** 3 Am. **St. Rep.,** 603.

111. Owners of ditches or **canals as** such, are
carriers and quasi-public servants.   **They** are awarded
certain privileges and are charged with certain duties
and subject to reasonable control.   Wheeler v. Northern
Colo. Irr. Co., 10 Colo., 582; S. C., 3 Am. St. Rep , 603;
see Const., art. XVI., sec. 6, note 57, etc.

EASEMENTS.

113. See Const., art. XVI., sec. 7, notes 4, etc.

114. Right of appropriator to enter the lands of another to make his appropriation effectual, see sec. 2264, note 2.

115. All lands in this territory are held in subordination to the the dominant right of others who must necessarily pass over them to obtain a supply of water to irrigate their lands. It is not, therefore, necessary that there should be a conveyance in writing to establish an easement for right of way for a ditch. Yunker v. Nichols, 1 Colo., 551.

116. The rule that the owner of a tract of land has an easement in a lower adjacent tract to the extent of burdening it with the water naturally flowing to it from the upper tract, applies only to waters naturally so flowing, and the servitude of the lower tract cannot be made more burdensome by the acts or industry of man. See Const., art. VI., sec. 7, note 4, etc.

EMINENT DOMAIN.

120. Statutes confirming the power to condemn private property to the use of another without the consent of the owner are in derogation of the common law, and must be strictly construed; see Const., art. XVI., sec. 7, notes; also art. II., secs. 14, 15, and notes.

124. The proprietor of an irrigating ditch has a property ownership both in the ditch and the right of way therefor, which cannot be taken or damaged for public use except upon payment of just compensation; Tripp v. Overocker, 7 Colo., 72.

MINERS' RIGHT TO USE OF WATER.

138. The use of water for mining purposes is one of the uses recognized and protected by the laws, both of the state and federal governments, and while an appropriator of the water of a natural stream is entitled to have such water flow down to him undiminished in quantity and any deterioration in quality, occasioned by

the use of it above for mining purposes, must be an injury without consequent damages; see People ex. rel., Wolpert v. Rogers et al., 12 Colo., 281, 1888.

REMEDIES—EQUITABLE.

150.   An action in equity lies for flooding when no laches in bringing suit.   Fuller v. Swan River Placer Min. Co., 12 Colo., 12 (1888).

152.   An appropriator of water is entitled to protection against acts which materially diminish the quantity of water to which he is entitled, or deteriorate its quality for use to which he desires to apply it.   Equity affords the appropriate remedy by way of an injunction for such wrongs.   Schilling v. Rominger, 4 Colo., 100.

REMEDIES—LEGAL.

168.   A person holding an assignment of shares of stock in a joint stock ditch company, but not transferred on the books of the company, is not entitled to waters from a ditch for the irrigation of his lands, not having used water therefrom, and if he take water by force from the ditch he is liable in trespass.   Supply Ditch Co. v. Elliott, 8 Colo., 330–335 (1887).   As to right to take water without knowledge of the ditch company see Coffin v. Left-Hand Ditch Co., 6 Id., 444–445 (1882).

172.   In an action for diversion of water the defendant is not liable for damages to another appropriator from the same stream resulting from a deficiency of the water supply, unless he was diverting from the stream more water than he was entitled to at the precise time the deficiency complained of existed.   Brown v. Smith, 10 Colo., 508.

RIPARIAN OWNERSHIP AND RIGHTS.

196.   Each riparian owner has a right, within his territory, to the use of the water as it flows, returning it to the channel of the river for the use of those below. Mason v. Cotton, 4 Fed. Rep., 792; 2 McCrary (Colo.), 82 (1880).

197. The remedy for the violation of the riparian rights is by action at law, and while equity may take cognizance of the violation of these rights, when conceded and established, it will not aid one who, out of mere captiousness, refuses to use water after it has been diverted from the stream by another, if he may so use it with substantially the same results as if obtained by continuous flow from the stream through his own race. Id.

STATUTE OF FRAUDS.

234. The right to conduct water over the lands of another is an interest in lands. See sec. 2019, notes; Yunker v. Nichols, 1 Colo., 552–557, etc. (1872); Schilling v. Rominger, 4 Id., 105 (1878). But it may be passed otherwise than by deed. Id., 109; Yunker v. Nichols, 1.Id., 570 (1872).

235. As to when a contract to appropriate water is not reached by this statute see Schilling v. Rominger, 4 Colo., 104 (1878); see also Whitsett v. Kershow, Id., 423.

Section 2257. Right of way through farms and lands. When any person owning claims in such locality has not sufficient length of area exposed to said stream to obtain a sufficient fall of water to irrigate his land, or that his farm, or land used by him for agricultural purposes, is too far removed from said stream, and that he has no water facilities on those lands, he shall be entitled to a right of way through the farms or tracts of land which lie between him and said stream, or the the farms or tracts of lands which lie above and below him on said stream, for the purposes hereinbefore stated. L. '61, p. 67, sec. 2; R. S. '68, p. 362, sec. 2; G. L. '77, p. 515, sec. 1373; G. S. '83, p. 561, sec. 1712.

1. For proceedings to condemn right of way, see chapter "Eminent Domain."

2. · Exemption of ditches from taxation. See sec. 2397; also sec. 568, 2260; also const., art. X., sec. 3.

3. See the case of Yunker v. Nichols, 1 Colo., 554, 566, 570 (1872). The points in this case are stated in

const., art. XVI., sec. 7, notes 4-14; Stewart v. Stephens, 10 Colo., 445 (1887).   See same notes.

4.   This section gives right of way for ditch, but it must be paid for.   Tripp v. Overocker, 7 Colo., 73 (1883); Downing v. Moore, 12 Id., 319 (1888).

5.   Instance where a bond in the sum of $10,000 was given, conditioned to pay the damages to be awarded for condemning right of way for ditch.   Davis v. Wannamaker, 2 Colo., 637 (1875).

6.   An order of court requiring defendant to build sluices for irrigating water, whenever necessary, is ineffectual for any purpose on account of its uncertainty. McKensie v. Ballard, 24 Pac. Rep., 1; 14 Colo.

Section 2258.   Extent of right of way.   Such right of way shall extend only to a ditch, dyke or cutting sufficient for the purpose required.   L. '81, p. 67, sec. 3; R. S. '68, p. 363, sec. 3; G. L. '77, p. 515, sec. 1374; G. S. '83, p. 561, sec. 1713.

1.   This section gives right of way for ditch, but it must be paid for.   Tripp v. Overocker, 7 Colo., 73 (1883); see also, Downing v. Moore, 12 Id., 319 (1888); see const., art. II., secs. 14, 15 and notes, and art. XVI., sec. 7 and notes.

Section 2259.   This section is unconstitutional and in conflict with laws subsequently enacted.   (Note by state engineer.)

Section 1762 of G. S. does away with sec. 2259.

Section 2260.   Condemnation of right of way. Upon the refusal of the owners of tracts of land or lands through which said ditch is proposed to run, to allow of its passage through their property, the person or persons desiring to open such ditch may proceed to condemn and take the right of way therefor (under the provisions of chapter thirty-one of these laws concerning eminent domain).   G. L. '77, p. 516, sec. 1376; G. S. '83, p. 561, sec. 1715.

Section 2261.   No land burdened with more than one ditch, except etc.   That no tract or parcel of

improved or occupied land in this state, shall, without
the written consent of the owner thereof, be subjected
to the burden of two or more irrigating ditches constructed for the purpose of conveying water through
said property to lands adjoining or beyond the same,
when the same object can feasibly and practicably be
attained by uniting and conveying all water necessary
to be conveyed through such property in one ditch. L.
'81, p. 164, sec. 1; G. S. '83, p. 562, sec. 1716.

1. It was competent for the legislature to adopt
this section; Tripp v. Overocker, et al., 7 Colo., 73,
1883. The constitutional provision granting the right
of way for the construction of ditches must be exercised
in such a way as to inflict the least possible inconvenience and injury upon the owner of the servient estate;
Id.

2. By this section a party cannot take a second
ditch across cultivated lands to irrigate his lands beyond,
when he can feasibly convey water through defendant's
ditch; Id.

3. But defendant's ditch must be one in which
there are rights of others which make it a burden to the
land; Downing v. Moore, 12 Colo., 319, (1888).

4. This section does not give the right to condemn
an enlargement of a mere private ditch on defendant's
own land, and not passing entirely through the same;
for such ditch is wholly at the will of the owner of the
land and constitutes no burden thereon; Tripp v. Overocker, *supra*. So far as intimating otherwise, is modified. See sec. 2263 and notes; Downing v. Moore, 12
Colo., 312, (1888).

Section 2262. Shortest route must be taken.
Whenever any person or persons find it necessary to convey water for the purpose of irrigation through the
improved or occupied lands of another, he or they shall
select for the line of such ditch through such property
the shortest and most direct route practicable upon
which said ditch can be constructed with uniform or
nearly uniform grade, and discharge the water at a point
where it can be conveyed to and used upon the land or

lands of the person or persons constructing such ditch. L. '81, p. 164, sec. 2; G. S. '83, p. 562, sec. 1717.

Section 2263. Owner of ditch must permit others to enlarge—Conditions. No person or persons having constructed a private ditch for the purposes and in the manner hereinbefore provided, shall prohibit or prevent any other person or persons from enlarging or using any ditch by him or them constructed in common with him or them, upon payment to him or them of a reasonable proportion of the cost of construction of said ditch. L. '81, p. 164, sec. 3; G. S. '83, p. 562, sec. 1718.

1. In so far as this section undertakes to limit or direct the compensation to be paid for the property, it is clearly unconstitutional and void. Tripp et al. v. Overocker et al., 7 Colo., 74 (1883).

2. Using or enlarging a ditch without the owner's consent is as much a taking or damaging of private property within the meaning of the constitution as would be appropriating the right of way therefor in the first instance. Id.

3. The right to enlarge and use the ditch of another already constructed will be enforced in the same manner, and under the same law as the right to take or damage any other kind of private property. Id.

4. But such ditch to be so enlarged hereunder must be such as to be a burden to the land, passing entirely through it, and not wholly and absolutely subject to the will of the owner of the land in being merely a private ditch thereon. So far as the case of Tripp v. Overocker, *supra*, holds otherwise, it is modified. See sec. 2261 and notes; Downing v. More, 12 Colo., 321 (1888).

Section 2264. When heads of ditches may be extended up stream—Condemnation—Proviso as to other ditches, etc. In case the channel of any natural stream shall become so cut out, lowered, turned aside or otherwise changed from any cause, as to prevent any ditch, canal, or feeder of any reservoir from receiving the proper inflow of water to which it may be entitled from such natural stream, the owner or owners of said ditch, canal or feeder shall have the right to extend the head of said ditch, canal or feeder to such distance up the stream which supplies the same as may be necessary for securing a sufficient flow of water into the same, and for that purpose shall have the same right to maintain proceedings for condemnation of right of way for such extension as in case of constructing a new ditch, and the priority of right to take water from such stream, through such ditch, canal or feeder as to any such ditch, canal or feeder shall remain unaffected in any respect by reason of such extension; *Provided*, However, that no such extension shall interfere with the complete use or enjoyment of any other ditch, canal or feeder. L. '81, pp. 161, 162, sec. 1; G. S. '83, p. 562, sec. 1719.

1. For condemnation proceedings, see chap. 45 "Eminent Domain;" Const., art. XVI., sec. 7.

2. An appropriation of water at a given point carries with it an implied authority to do all that shall become necessary to secure the benefit of such appropriation. To this extent the appropriator acquires an easement in the adjoining lands. This right is, however, restricted to the narrowest limits, and it must be exercised in such manner as to occasion as little damage as possible to the owner of the adjoining premises. Crisman v Heiderer, 5 Colo., 589; see sec. 2256, note.

Section 2265. Ditch owners must file map and statement—Priority. Every person, association or corporation hereafter constructing or enlarging any ditch, canal or feeder for any ditch or reservoir for irrigation and taking water directly from any natural stream, and of a carrying capacity of more than one cubic foot of water per second of time, as so constructed or enlarged, shall, within ninety (90) days after the commencement of such construction or enlargement, file in the office of the county clerk and recorder of the county in which the headgate of such ditch or feeder may be situated, and also in the office of the state hydraulic engineer, a map showing the point of location of such headgate; the route of such ditch or canal, or the high water line of such reservoir or reservoirs, and the route of the feeder to, and ditches or canals from, such reservoir or reservoirs; the legal subdivisions of the lands upon which such structures are built, or to be built, if on surveyed lands; the names of the owners of such lands, as far as the same are of record in the office of the county clerk of the county in which they are situated; such courses, distances and corners, by reference to legal subdivisions, if on surveyed lands, or to natural objects if on unsurveyed lands, as will clearly designate the location of such structures. Upon or attached to such map shall be a statement showing :

First—The point of location of the headgate above mentioned.

Second—The depth, width and grade of such ditch, canal or feeder.

Third—The carrying capacity of such ditch, canal or feeder, in cubic feet per second of time, and the capacity of such reservoir or reservoirs in cubic feet, when filled to the high water mark.

Fourth—The time of commencement of work on such structures, which time may be dated from the commencement of the surveys therefor. In case of an enlargement, such statement shall also show the matters required in items second, third and fourth above, as to the enlargement, and state the increased capacity arising from such enlargement. If such statement be filed within the time above limited, priority of right of way, and water accordingly, shall date from the day named as the day of commencing work; otherwise, only from the date of the filing of the same; *Provided*, That nothing herein contained shall be taken to dispense with the necessity of due diligence in the prosecution of such structures on the part of the projectors of the same. Such statement shall be signed by the person, association or corporation on whose behalf it is made, and the truth of the matters shown in such map and statement shall be sworn to by some person in whose personal knowledge the truth of the same shall lie. L. '81, p. 162, sec. 2; G. S. '83, pp. 562, 563, sec. 1720; amended, L. '87, p. 314, etc., sec. 2.

1. Filing **statement of claim, etc. See sec.** 2400, also 2424.

2. Organization of corporation. See sec. 472, **570.**

3. Ditch companies additional matter. **See sec.** 567, 575.

4. **Requirements of** ditch owners. **See sec. 2287-** 2293.

5. **Duty of state engineer.** See sec. 2460.

6. **This section, as it** formerly stood, **referred to in** Crisman **v. Heiderer, 5 Colo., 594** (1881).

Section 2266. Only irrigation ditches referred to in the **last** above **section. This act shall** apply to and affect **only** ditches, **canals or feeders used** for carrying water **for** the **purpose of irrigation, and** for no other purpose whatever. **L. '81, p. 161, sec. 3;** G. S. '83, p. 563, **sec.** 1721.

1. This section referred **to in** Crisman v. Heiderer, **5 Colo.,** 594 (1881).

Section 2267. **Water** to be **equally** divided **among** consumers pro-rated. **If** at any time any ditch **or reser-** voir from which water **is** or shall be drawn for **irriga-** tion shall not be **entitled to** a full supply of water **from** the natural stream **which** supplies the **same, the water** actually **received into and** carried by **such ditch, or held** in such **reservoir, shall be** divided **among all consumers** of water **from such ditch or reservoir, as well as the** owners, **shareholders or** stockholders **thereof,** as the par- ties purchasing **water** therefrom, **and** parties taking water partly under **and by** virtue of holding shares and partly by purchasing **the** same, **to** each his share *pro rata*, according to the amount he, she or **they** (in cases in which several consume water jointly) **shall** be then entitled, so that all owners and purchasers shall suffer **from** the deficiency arising **from** the cause aforesaid **each in** proportion to the amount of water **to** which he, **she or** they should have received in case **no** deficiency **of water had** occurred. L. '79, p. 97, sec. 4; G. S. '83, **p.** 563, **sec. 1722;** see L. '61, p. 68, sec. 4.

1. As to pro-rating, under L. '61, p. 68, sec. 4, see Coffin v. Left-Hand Ditch Co., 6 Colo., 448 (1882).

2. As to pro-rating, under this section, see F. H. L. C. & R. Co. v. Southworth, 13 Colo., 123-135 (1889); fully set out in Const., art. XVI., sec. 6, notes 93, etc.

3. Time when **water shall** flow in irrigating ditches. Sec. 2287.

Section 2268. Irrigation of meadows—Right **to make ditch**—Priority. All persons who shall have enjoyed the use of the water in any natural stream for the irrigation of any meadow land, by the natural overflow or operation of the water of such stream, shall, in case the diminishing of water supplied by such stream, for any cause, prevent such irrigation therefrom in as ample a manner as formerly, have right to construct a ditch for the irrigation of such meadow, and to take water from such stream therefor, and his or their right to water through such ditch shall have the same priority as though said ditch had been constructed at the time he, she or they first occupied and used such land as meadow ground. L. '79, p 106, sec. 37; G. S. '83, pp. 563, 564, sec. 1723.

Section 2269. **Priority of right.** That all ditches now constructed or hereafter to be constructed for the purpose of utilizing the waste, seepage or spring waters of the state, shall be governed by the same laws relating to priority of right as those ditches constructed for the purpose of utilizing the waters of running streams; *Provided*, That the person upon whose lands the seepage or spring waters first arise, shall have the prior right to such waters if capable of being used upon his lands. L. '89, p. 215, sec. 1.

Section 2270. Reservoirs—Right to water—Right of way—Condemnation—Embankments over ten feet submit to county board. Persons desirous to construct and maintain reservoirs for the purpose of storing water, shall have the right to take from any of the natural streams of the state and store away any unappropriated water not needed for immediate use for domestic or irrigating purposes ; to construct and maintain ditches

for carrying such water to and from such reservoir, and
to condemn lands for such reservoirs and ditches, in the
same manner provided by law for the condemnation of
land for right of way for ditches ; *Provided*, No reservoir with embankments or a dam exceeding ten feet in
height shall be made without first submitting the plans
thereof to the county commissioners of the county in
which it is situated, and obtaining their approval of
such plans. L. '79, pp. 106-107, sec. 38; G S. '83, p.
564, sec. 1724.

1. See section 2460.

2. The constitution and statute recognize the right
to construct and maintain reservoirs. Larimer Co. R.
Co. v. People, ex rel., 8 Colo., 615 (1885); see also sec.
2256, notes.

Section 2271. Conducting water in natural streams—
Taking out—Allowance for seepage—How determined.
The owners of any reservoir may conduct the water
therefrom into and along any of the natural streams of
the state, but not so as to raise the waters thereof above
ordinary high-water mark, and may take the same out
again at any point desired, without regard to the prior
rights of others to water from said stream ; but due
allowance shall be made for evaporation and scapage
(seepage), the amount to be determined by the commissioners of irrigation of the district ; or, if there are no
such commissioners, then by the county commissioners
of the county in which the water shall be taken out for
use. L. '79, p. 107, sec 39; G. S. '83, p. 564, sec. 1725.

Section 2272. Liability of owners for damages—
The owners of the reservoirs shall be liable for all damages arising from leakage or overflow of the waters
therefrom, or by floods caused by breaking of the
embankments of such reservoirs. L. '79, p. 107, sec. 40;
G. S. '83, p. 564, sec. 1726.

1. As to negligence and right of action without
proving same, see decisions under sec. 3713 ; see also
sec. 571, and notes and secs. 2274, 2282 and notes.

2. No dam to overflow roads, sec. 3961.

Section 2273.   Right to place wheels on streams—
Conditions.   All persons on the margin, brink, neighborhood or precinct of any stream of water, shall have
the right and power to place upon the bank of said
stream a wheel, or other machine for the purpose of
raising water to the level required for the purpose of irrigation, and the right of way shall not be refused by the
owner of any tract of land upon which it is required,
subject, of course, to the like regulations, as required for
ditches, and laid down in sections hereinbefore enumerated.   L. '61, pp. 68, 69, sec. 8; R. S. '68, p. 364, sec. 6;
G. L. '77, p. 516, sec. 1377; G. S. '83, p. 564, sec. 1727.

DIVISION II.

DUTIES OF OWNERS.

Section 2274.   Owner shall maintain embankments
—Tail ditch.   The owner or owners of any ditch for
irrigation or other purposes shall carefully maintain
the embankments thereof, so that the waters of such
ditch may not flood or damage the premises of others,
and shall make a tail ditch, so as to return the water in
such ditch with as little waste as possible into the
stream from which it was taken.   R. S. '68, p. 364, sec.
7; amended L. '72, p. 144, sec. 1; L. '76, p. 78, sec. 2;
G. S. '83, pp. 564, 565, sec. 1728.

When water to be kept flowing and how ditches to
be kept in repair, sec. 2287, 2288; see also sec. 571 anti;
royalty, bonus, etc., sec. 2304; see as to the liability of
owner of ditch secs. 2272, 2282 and notes.   When negligence need not be proved, see sec. 2272 and note and
sec. 3713, note.

2.   This section with sec. 2278 imposes upon the
owner of every ditch the duty to keep the ditch in such
good condition and repair that the water from the same
cannot readily escape therefrom to the injury of any
property.   Greeley Irr. Co. v. House, 24 Pac. Rep., 330
(1890), 14 Colo.

3. A ditch company is liable for damages caused by allowing the water of a ditch to overflow the lands of others. Id.

4. The liability of ditch owners under this section for damages caused by ditches overflowing, arises from their failure to exercise ordinary care in preventing the escape of the water. Id.

5. A remedy may be had in damages for injuries resulting from an exercise of lawful powers in an improper, careless or negligent manner. Id., Ditch Co. v. Anderson, 8 Colo., 143 (1884); Water Co. v. Middaugh, 12 Id., 440, 1889.

6 Ditch owners who are grossly negligent in keeping their ditches in repair, cannot be permitted to take refuge under the plea of unavoidable accident. Greeley Irr. Co. v. House, 24 Pac. Rep., 331 (1890), 14 Colo.

7. The owners of a ditch who permit the water to overflow the banks of the ditch and flood the lands of another, where they had been warned that the ditch was running too full and that the water was in danger of escaping unless the flow was diminished, are liable for all damages so caused. Id.

Section 2275. Vested rights of mill and ditch owners. Nothing in this chapter contained shall be so construed as to impair the prior vested rights of any mill or ditch owner or other person to use the waters of any such water course. L. '61, p. 69, sec. 10; amended R. S. '68, p. 364, sec. 8; G. S. '77, p. 516, sec. 1379; G. S. '83, p. 565, sec. 1729.

1. See sec. 571 and notes.

Section 2276. Crossing highways—Bridge. Any ditch company constructing a ditch, or any individual having ditches for irrigation, or for other purposes, whenever the same be taken across any public highway or public traveled road, shall put a good substantial bridge, not less than fourteen feet in breadth, over such watercourse where it crosses said road. R. S. '68, p. 364, sec. 10; G. L. '77, pp. 516, 517, sec. 1381; G. S. '83, p. 565, sec. 1730.

1. Canals to be covered when. Sec. 2278.

Section 2277. Ditch must be bridged in three days —Duty of supervisor. When any such ditch or water-course shall be constructed across any public traveled road, and not bridged in three days thereafter, it shall be the duty of the supervisor of the road district to put a bridge over said ditch or water-course, of the dimensions specified in section ten of this chapter, and call on the owner or owners of the ditch to pay the expenses of con-structing such bridge. R. S. '68, p. 364, sec. 11; G. L. '77, p. 517, sec. 1382; G. S. '83, p. 565, sec. 1731.

Section 10 referred to in last above section is section 2276 of this chapter.

2. Supervisor is road overseer. Secs. 3952 and 3962.

Section 2278. Canals and ditches—When and by whom covered. That every corporation and company, whether created by special act or organized under the general incorporation laws of this state, and every part-nership, person or persons who now, or may at any time hereafter, own or control any canal or ditch, or any part thereof, being two feet in width or over, and carrying water to the depth of twelve inches or over, which ca-nal or ditch, or any part thereof, is within the corporate limits of any city denominated in the law as of first class, or any city existing by special charter of a popula-tion equal to or exceeding said cities of first class, or any of the additions thereto, shall, at their own expense, within sixty days after this act shall have taken effect, confine, flume and cover over all or any part of such ca-nal or ditch, whether located on or across private prop-erty, public highways or alleys in said city or additions thereto, in a reasonable and sufficient manner, and with such materials as will render such fluming or covering safe and a sure protection to the lives and property of the inhabitants of that city; and any such corporation, company, partnership, person or persons, shall at all times thereafter keep and maintain any and all such structures confining, fluming and covering of such canal or ditch in good order and repair, at their own expense. L. '87, p. 65, sec. 1.

Section 2279. Head to be latticed. Such corporation, company, partnership, person or persons, shall, at their own expense, safely and securely lattice or slat the head of such flume or covering with proper materials, so that persons or animals cannot accidentally enter such flume or covering at the head thereof, and pass or be carried down the current of said canal or ditch, and shall thereafter maintain and keep the same in good order and repair at their own cost and expense. L. '87, p. 66, sec. 2.

Section 2280. Penalty for failure or refusal—Proviso. If any such corporation, company, partnership, person or persons, shall fail or refuse to comply with any of the provisions of the two preceding sections, such corporation, company, partnership, person or persons, shall forfeit and pay to the county, for the use of the common school fund, the sum of fifty dollars for each and every day such failure or refusal shall continue, to be recovered by a civil action in the name of the people of the state of Colorado in any court of competent jurisdiction; *Provided*, That nothing in this act shall be construed to bar an action for special damages by any person who shall have suffered such damages by reason of any failure to comply with any of the provisions of this act. L. '87, p. 66, sec. 3.

Section 2281. Proceedings against owner for payment—Damages. If the owner or owners of such ditch refuse to pay the bill of expenses so presented the supervisor may go before any justice of the peace in the township or precinct, and make oath to the correctness of the bill and that the owner or owners of the ditch refuse payment; and thereupon such justice of the peace shall issue a summons against such owner or owners, requiring him or them to appear and answer to the complaint of such supervisor in an action of debt for the amount sworn to be due, such summons to be made returnable and served, and proceedings to be had thereon as in other cases; and in case judgment shall be

given against such owner or owners the justice shall
assess, in addition to the amount sworn to be due as
aforesaid, the sum of ten dollars, as damages arising
from the delay of such owner or owners, such judgment
to be collected as in other cases, and to be a fund in the
hands of the supervisor of roads, for the repair of roads
in such precinct or district. R. S. '68, p. 365, sec. 12;
G. L. '77, p. 517, sec. 1383; G. S. '83, p. 565,
sec. 1732.

1. See "Roads and Highways," chap. 108.

2. The Supervisor, secs. 3952, 3962.

Section 2282. Owner of ditch must prevent waste.
The owner of any irrigating or mill ditch shall carefully maintain and keep the embankments thereof in
good repair, and prevent the water from wasting. L.
'76, p. 78, sec. 1; G. L. '77, p. 518, sec. 1385; G. S.
'83, p. 565, sec. 1733.

1. See secs. 2272, 2274, notes.

2. See Greeley Irr. Co. v. House, 24 Pac. Rep.,
330, (1890), 14 Colo., in notes to secs. 571, 2274, these
statutes.

Section 2283. Running excess of water forbidden.
During the summer season it shall not be lawful for
any person or persons to run through his or their irrigating ditch any greater quantity of water than is abso
lutely necessary for irrigating his or their said land,
and for domestic and stock purposes; it being the intent
and meaning of this section to prevent the wasting and
useless discharge and running away of water. L. 76,
p. 78, sec. 2; G. L. '77, p. 518, sec. 1386; G. S. '83,
p. 566, sec. 1734.

Section 2284. Penalty for violation of this act.
Any person who shall wilfully violate any of the provisions of this act shall, on conviction thereof before
any court having competent jurisdiction, be fined·in a
sum of not less than one hundred (100) dollars. Suits
for penalties under this act shall be brought in the name
of the people of the state of Colorado. L. '76, p. 78,

sec. 3; G. L. '77, p. 518, sec. 1387; G. S. '83, p. 566, sec. 1735.

1. See "Penalties," sec. 3963.

Section 2285. Owners keep headgate.—Size of timbers. That the owner or owners of every irrigating ditch, flume or canal, in this state shall be required to erect and keep in good repair a headgate at the head of their ditch, flume or canal. Such headgate, together with the necessary embankments, shall be of sufficient height and strength to control the water at all ordinary stages. The frame work of such headgate shall be contructed of timber not less than four inches square, and the bottom, sides, and gate or gates, shall be of plank not less than two inches in thickness. L. '81, p. 165, sec. 1; G. S. '83, p. 566, sec. 1736.

Section 2286. Liability of owner for neglect, refusal. Owners of all ditches shall be liable for all damages resulting from their neglect or refusal to comply with the provisions of section one of this act. L. '81, p. 165, sec. 2; G. S. '83, p. 566, sec. 1737.

1. Section one referred to in this section is section 2285 hereof.

Section 2287. When water shall be kept flowing in ditches—Proviso. Every person or company owning or controlling any canal or ditch used for the purposes of irrigation shall, during the time from April 15 to November 1, in each year, keep a flow of water therein, so far as may be reasonably practicable for the purpose of irrigation, sufficient to meet the requirements of all such persons as are properly entitled to the use of water therefrom, to the extent, if necessary, to which such person may be entitled to water, and no more ; *Provided, however,* That whenever the rivers, or public streams, or sources from which water is obtained, are not sufficiently free from ice, or the volume of water therein is too low and inadequate for that purpose, then such canal or ditch shall be kept with as full a flow of water therein as may be practicable, subject, however, to the rights of priorities from the streams or other sources, as provided by law, and the necessity of cleaning, repairing and maintaining the same in good condition   L. '87, p. 304, sec. 1.

Section 2288.   Ditches are to be kept in repair—Outlets.   The owners or persons in control of any canal or ditch used for irrigating purposes shall maintain the same in good order and repair, ready to receive water by April 15, in each year, so far as can be accomplished by the exercise of reasonable care and diligence, and shall construct the necessary outlets in the banks of the canal or ditch for a proper delivery of the water to persons having paid-up shares, or who have rights to the use of water; *Provided, however*, That a multiplicity of outlets in the canal or ditch shall at all times be avoided, so far as the same shall be reasonably practicable, and the location of the same shall be under the control of, and shall be at the most convenient and practicable points consistent with the protection and safety of the ditch for the distribution of water among the various claimants thereof ; and such location shall be under the control of a superintendent.   L. '87, p. 305, sec. 2.

Section 2289.   Superintendent to measure water. It shall be the duty of those owning or controlling such canals or ditches, to appoint a superintendent, whose duty it shall be to measure the water from such canal or ditch through the outlets, to those entitled thereto according to his or her *pro rata* share.   L. '87, p. 305, sec. 3.

Section 2290.   Penalty for refusal or neglect to deliver water.   Any superintendent, or any person having charge of the said ditch, who shall wilfully neglect or refuse to deliver water, as in this act provided, or any person or persons who shall prevent or interfere with the proper delivery of water to the person or persons having the right thereto, shall be guilty of a misdemeanor, and upon conviction thereof shall be subject to a fine of not less than ten nor more than one hundred dollars for each offense, or imprisonment not exceeding one month, or by both such fine and imprisonment; and the money thus collected shall be paid into the general fund of the county in which the misdemeanor has been committed; and **the owner or owners of such** ditches shall be liable in damages **to the** person or persons deprived of the use of **the water to** which they **were entitled,** as in this act provided. **L. '87,** p. 305, **sec. 4.**

Section 2291. Water commissioner **to promptly** measure water. Any water commissioner, **or his deputy,** or assistant, who shall wilfully neglect or **refuse,** after being called upon in accordance with section **1758** of the general statutes **of** the state, to promptly measure water from the stream, or other source of supply, into the irrigating canals **or** ditches, **in** his district, according to their respective **priorities, to** the extent to which water **may be** actually **necessary for the** irrigation of lands **under such canals or ditches, shall** be deemed guilty **of a** misdemeanor, **and shall be** subject to the same penalty **as** provided **in sec. 4 of this** act. L. '87, **p. 305,** sec. **5.**

1. **Sec.** 4 referred **to is sec. 2290** hereof.

2. Sec. 1758 referred **to is sec. 2392** hereof.

Section. 2292. Duties of justices of the peace. In all cases declared misdemeanors by this act, any justice of the peace of the county in which the offense was committed, may, upon complaint being made, as **is** now required by law, issue a warrant directed to any proper officer of the county for the arrest of any person so charged with any such misdemeanor, and upon the arrest of such person or persons, the justice of the peace before whom such person or persons may be brought for trial, shall hear and determine the cause, and, if he find the accused guilty, shall assess the fine, and if imprisonment be a portion of the sentence, then to fix the term of imprisonment, or both, as provided in section 4 of this act; *Provided,* The accused may have a trial by jury which shall be summoned as in cases before justices of the peace for assault **and** battery. L. '87, p. 306, sec. 6.

Section 2292.

1. **Sec. 4** referred to is sec. 2290 hereof.

Section 2293. Erect and maintain headgates. All persons, associations or corporations who have heretofore or who may hereafter divert water for purposes of irrigation from any of the public streams of the state shall erect and maintain headgates and wastegates in connection therewith, and in case of failure or neglect or refusal to do so after five days' notice has been given by the water commissioner or state engineer, then said headgates shall be constructed by the water commissioner of the district within which said ditch, canal or conduit may be located, and if, upon demand, the owner or owners of said ditch, canal or conduit shall neglect or refuse to pay the expenses thereof, then the said water commissioner shall take such proceedings to recover the same as is now provided for by sections 1730, 1731 and 1732 of the General Statutes of 1883 in the case of failure to build and maintain bridges. L. '89, p. 161, sec. 1.

1. Secs. 1730, 1731 and 1732, referred to, are secs. 2276, 2277 and 2281, respectively hereof.

Section 2294. Keep suitable locks and fastenings on headgates. All persons, associations or corporations shall put and keep suitable locks and fastenings on their headgates, where water is conducted from the public streams or heads of supply, and if said persons, associations or corporations refuse or neglect to provide locks and suitable fastenings for said headgates after five days' notice by the water commissioner of the district, or by the state enginer, it is made the duty of the water commissioner of the water district and its superintendent to provide suitable locks and fastenings, and if the owner or owners of said ditch, canal or conduit shall neglect or refuse to pay the expenses thereof the water commissioner shall take such proceedings to recover the same as are provided in sec. 1 of this act, the keys of said locks to be under the control and in the possession of the water commissioner of the district during the season of irrigation or domestic distribution of water. L. '89, p. 161, sec. 2.

1. Sec. 1, referred to, is sec. 2293 hereof.

DIVISION III.

RATES OF CHARGES FOR WATER.

Section 2295. Regulating charges—Petition—Affidavits — Proceedings before commissioners—Notice—Service—Evidence—Depositions. The county commissioners of each county shall, at their regular January session in each year, hear and consider any and all applications which may be made to them by any party or parties interested in procuring water for irrigation by purchase from any ditch or reservoir furnishing and selling water, or proposing to furnish water for sale, the whole or upper part of which shall lie in such county, which application shall be supported by such affidavit or affidavits as the applicant may see proper to present, showing reasonable cause for such board to proceed to fix the price of water to be thereafter sold from such ditch or reservoir ; and (if) such board of commissioners shall, upon examination of such affidavit or affidavits, or from the oaths of witnesses in addition thereto, find that the facts sworn to show the application to be in good faith, and that there is reasonable grounds to believe that unjust prices are, or are likely to be, charged for water from such ditch or reservoir, they shall enter an order fixing a day, not sooner than forty days thereafter, nor later than the third day of the (next) regular session of their board, when they will hear all parties directly or indirectly interested in said ditch or reservoir, or in procuring water therefrom for irrigation, who may appear, [as well as all testimony by witnesses, or depositions taken on notices as hereinafter provided touching the said ditch or reservoir, and the cost of furnishing water therefrom, at which time all persons or corporations interested in said ditch or reservoir, as well as all interested in obtaining water therefrom, or in lands which may be irrigated therefrom, may appear by themselves, their agents or attorneys, and said commissioners shall then proceed to take action in the matter of fixing such price of water, *Provided*, The applicant shall, within ten days from the time of entering such order, cause a

copy thereof, duly certified, to be delivered to the owner
of such ditch or reservoir, if it be owned by one person,
or each of the owners, if it be owned by several persons,
or to the president, secretary or treasurer of the company, if it belongs to a corporation or association having
such officers; or, if such owner cannot be found, he shall
cause such copy to be left at his usual place of residence
with some person or member of his family residing
there, and over fourteen years of age, and if such ditch
officer cannot be found, he shall cause such copy to be
left at the office or place of business of the company of
which he is such officer, or at his residence, if such company have no place of business, and if such ditch is
owned by several owners, not an incorporated company,
it shall be sufficient to serve such notice by delivering
one such copy each to a majority of them, and such applicant shall make affidavit of the manner in which such
copy or copies have been served. Depositions mentioned in section one hereof, to be used before said commissioners, shall be taken before any officer in the state
authorized by law to take depositions, upon reasonable
notice being given to the opposite party of the time
and place of taking such depositions. L. '79, pp. 94-
96, sec. 1; G. S. '83, pp. 566, 567, sec. 1738.

1. Fixing rates. Sec. 2298, etc.

2. Rates fixed by county commissioners. Sec. 570.

4. The title to this act is constitutional and em-
braces the fixing of water rates. Golden Canal Co. v.
bright, 8 Colo., 147, etc. (1884.)

4. The purpose and substance of this section is
constitutional. Id.

5. As to adjudicating water rights. See sec. 2399,
etc; also const., art. XVI., sec. 6, notes 48, 9 and 93,
etc.

Section 2296. Powers and duties of board—Subpoenas—Compulsion—Adjournments—Examinations—
Facts—Order—Proviso as to contracts. Said board shall
hear and examine all legal testimony or proofs offered by
any of the parties interested as before mentioned, as

well concerning the value of the construction of such ditch or reservoir as the cost and expense of maintaining and operating the same, and all matters which may affect the just price and value of water to be furnished therefrom; and they shall have power to issue subpoenas to witnesses and compel their attendance, which subpoenas shall be served by the sheriff of the proper county when required, and also to compel the production of books and papers required for evidence in as full and ample a manner as the district court now has. They may adjourn the hearing from time to time to further the ends of justice or suit the general convenience of parties. Upon hearing an [and] considering all the matters and facts involved in the case, the board of commissioners shall enter an order naming and describing the ditch or reservoir with sufficient certainty, and fixing a just price upon all water to be hereafter sold, which price shall not be thereafter changed oftener than once in two years; *Provided,* That no price so fixed shall effect [affect] the rights of the parties, or their lawful assignees or grantees, who may have contracts with the company, association or person owning such ditch or reservoir, or their lessees, grantees or successors, nor the rights of such owners, lessees or grantees under such contract, nor shall it in any way affect or hinder the making of such contract. L. '79, p. 96, sec. 2; G. S. '83, pp. 567, 568, sec. 1739.

1. When compelled to furnish water, rates, sec. 570. The constitution gives the county commissioners power to fix rates. Const., art. XVI., sec. 8.

2. The substance and purpose of this section is constitutional. Golden Canal Co. v. Bright, 8 Colo., 147 (1884).

3. There is no appeal from the decision of the commissioners. Id., 155.

Section **2297.** Right to continue purchasing water—Tender of price—Stockholders—Rights. Any person or persons, acting jointly or severally, who shall have purchased and used water for irrigation for lands occupied by him, her or them, from any ditch or reservoir, and shall not have ceased to do so for the purpose or with the intent to procure water from some other source of supply, shall have a right to continue to purchase water to the same amount for his, her or their lands, on paying or tendering the price thereof fixed by the county commissioners as above provided ; or, if no price shall have been fixed by them, the price at which the owners of such ditch or reservoir may be then selling water, or did sell water during the then last preceding year. This section shall not apply to the case of those who may have taken water as stockholders or shareholders after they shall have sold or forfeited their shares or stock, unless they shall have retained a right to procure such water by contract, agreement or understanding, and use between themselves and the owners of such ditch, and not then to the injury of other purchasers of water from or shareholders in (the) same ditch. L. '79, pp. 96, 97, sec. 3; G. S. '83, p. 568, sec. 1740.

1. When compelled to furnish, sec. 570; Const. Colo., art. XVI., sec. 8. Receiving money or other valuable thing, etc., as a prerequisite to granting water illegal; penalty, sec. 2304, etc. Superintendent for water division ; appointment of, etc., sec. 2447, etc.

2. This section confers a confirmative right upon the prior purchaser who has complied with the provisions thereof, to continue his purchase of water, and he cannot be required as a condition precedent to the exercise of this right, to acknowledge the equity of all rules adopted by the ditch company ; Golden Canal Co. v. Bright, 8 Colo., 149, (1884.)

3. Whether the statute imposes upon the ditch owner the duty of keeping sufficient water in the ditch, when possible, to supply prior purchasers, *quere*; Id., 152.

4. A prior purchaser is entitled to continue to purchase, although he may be able to obtain water from some other source. Id.

5. This section applies only to those parties who have exercised the right to use water for their lands, and is an assurance of the right to continue the use of the

water, and this right may be enforced by mandamus. It does not give one who has never had the use of water the right to the water. Wheeler v. North Colo. I. Co., 10 Colo., 595, (1887); Golden Canal Co. v. Bright, 8 Id., 152 (1884).

6. Instance where a shareholder had disposed of a portion of his share of stock. Supply Ditch Co. v. Elliott, 10 Colo., 328 (1887).

7. Where a party sues for damages caused by being restrained from using the water from a certain ditch, if it is shown that he could have obtained sufficient water from another source, he will not be entitled to receive a greater sum than he would have had to expend to obtain water from such source. Mack v. Jackson, 9 Colo., 537 (1886).

Section 2298. County commissioners hear and consider applications. The county commissioners of each county shall, at their regular sessions in each year, and at such other sessions as they in their discretion may deem proper, in view of the irrigation and harvesting season, and the convenience of all parties interested, hear and consider all applications which may be made to them by any party or parties interested either in furnishing and delivering for compensation in any manner, or in procuring for such compensation, water for irrigation, mining, milling, manufacturing, or domestic purposes, from any ditch, canal or conduit, or reservoir, the whole or any part of which shall lie in such county. Which application shall be supported by such affidavits as the applicant or applicants may present, showing reasonable cause for such board of county commissioners to proceed to fix a reasonable maximum rate of compensation for water to be thereafter delivered from such ditch, canal, conduit or reservoir within such county. L. '87, p. 291, sec. 1.

1. Prior to this act the statute did not authorize the county commissioners of a given county to establish a maximum rate if the head of the canal was located in in another county; Wheeler v. North Col. Irr. Co., 10 Colo., 583, (1887).

Section 2299. Commissioners appoint day for hearing parties interested in ditches, etc. Every such board of commissioners shall upon examination of such affidavit or affidavits, or from the oaths of witnesses in addition thereto, if they find that the facts sworn to show the application to be in good faith, and that there are reasonable grounds to believe that unjust rates of compensation are, or are likely to be, charged or demanded for water from such ditch, canal, conduit or reservoir, shall enter an order fixing a day not sooner than twenty days thereafter, nor later than the third day of the next regular session of their board, when they will hear all parties interested in such ditch or other water-works as aforesaid, or in procuring water therefrom, for any of the said uses, as well as all documentary or oral evidence or depositions, taken according to law, touching the said ditch, or other work as aforesaid, and the cost of furnishing water therefrom. L. '87, p. 292, sec. 2.

Section 2300. Commissioners fix rates—Proviso. At the time so fixed all persons interested, as aforesaid, on either side of the controversy in lands which may be irrigated from such ditch, or other work aforesaid, may appear by themselves, their agents or attorneys, and said commissioners shall then proceed to take action in the matter of fixing such rates of compensation for the delivery of water; *Provided*, The applicant or applicants (if the application be made by a party or parties, as aforesaid, desirous of procuring water) shall, within ten days from the time of entering the said order fixing the hearing, cause a copy of such order, duly certified, to be delivered to the owner or owners of such ditch, canal, conduit or reservoir, or to the president, secretary or treasurer of the company, if it be owned by a corporation or association having such officers. If any such owner cannot be found a copy shall be left at his usual place of abode, with some person residing there over twelve years of age; and if such officer of any corporation or association cannot be found such copy shall be left at the usual place of business of the company of which he is such officer, or at his residence if such company have no place of business; and if such ditch, or other work aforesaid, **shall be owned** by several **owners** not being an incorporated company **it** shall **be** sufficient to serve such notice by delivering copies to a majority of them. If the applicant be the owner or party **controlling** such ditch, canal, conduit or reservoir, such notice shall be given by causing printed copies of such **order, in** hand-bill form, in conspicuous type, **to be posted securely in** ten or more public places throughout **the** district watered from such ditch, or other work aforesaid (if the water be used for irrigation), and **one** copy shall be posted **for** every mile **in** length of such ditch; but if such ditch or other work be for the supply of **water for** milling **or mining** it **shall** be sufficient to serve such **copy on** the **parties** then taking **water** therefrom. **The person or persons making** such **service or posting such printed copies shall make** affidavit of the **manner in which the same has** been done, which affi**davit shall be filed** with the **said board** of county com**missioners.** Depositions mentioned in sec. **2** hereof, **to be used** before said commissioners, **shall** be taken **before** any **officer in** the state authorized **by law** to take depositions, upon reasonable notice being given to the **opposite** party of the time and place of taking the **same.** L. '87, p. 292, sec. **3.**

1. Sec. 2, referred **to in** this section, is sec. 2299 hereof.

Section 2301. Commissioners **may postpone hear**ing—Witnesses—Subpoenas—Court **compel obedience.** Said **board** of commissioners may **adjourn or postpone any** hearing from time to time as **may be found neces**sary, **or** for the convenience of **parties, or of public** business; and **they shall** hear and **examine all legal tes**timony or proofs offered by any party interested, as afore**said,** as well concerning the original **cost and present** value of works **and structure** of such ditch, canal, conduit or reservoir, **as the cost** and expense of maintaining and operating the **same, and** all matters **which** may affect the establishing **of** a reasonable maximum rate of compensation for water to be furnished and delivered therefrom; **and** they may **issue** subpoenas **for** witnesses, which subpoenas shall be **served** by the sheriff of the county, who shall receive the lawful fees for all such service ; and said board may also issue a subpœna for the production of all books and papers required for evidence before them. Upon hearing and considering all the evidence and facts, and matters involved in the case, said board of commissioners shall enter an order describing the ditch, canal, conduit, reservoir or other work in question, with sufficient certainty, and fixing a just and reasonable maximum rate of compensation for water to be thereafter delivered from such ditch or other work as last aforesaid, within the county in which such commissioners act; and such rate shall not be changed within two years from the time when they shall so be fixed, unless upon good cause shown. The district court of the proper county, or the judge thereof in vacation, may, in case of refusal to obey the subpœnas of the board of county commissioners, compel obedience thereto, or punish for refusal to obey, after hearing, as in cases of attachment for contempt of such district court. L. '87, p. 293, sec. 4.

Section 2302. False swearing. Every person who shall swear or affirm falsely in any manner, or testify falsely after being duly sworn or having affirmed as a witness in any proceeding provided for in this act, shall be deemed guilty of perjury, and on conviction shall be punished accordingly. L. '87, p. 294, sec. 5.

Section 2303. Repeal. All acts and parts of acts inconsistent with the provisions of this act are hereby repealed; but such repeal shall not work any interference with any proceeding by any board of county commissioners now pending, saving that any such proceeding may, at the request of either party, be carried on to completion under the provisions hereof. L. '87, p. 294, sec. 6.

Section 2304. Royalties prohibited—Illegal rate— Excess recovered—Costs. It shall not be lawful for any person owning or controlling, or claiming to own or control any ditch, canal or reservoir, carrying or storing, or designed for the carrying or storing of any water taken from any natural stream or lake within this state, to be furnished or delivered for compensation, for irrigation, mining, milling or domestic purposes, to persons not interested in such ownership or control, to demand, bargain for, accept or receive from any person who may apply for water for any of the aforesaid purposes, any money or other valuable thing whatsoever, or any promise or agreement therefor, directly or indirectly, as royalty, bonus or premium prerequisite or condition precedent to the right or privilege of applying, or bargaining for, or procuring such water. But such water shall be furnished, carried and delivered upon payment or tender of the charges fixed by the county commissioners of the proper county, as is, or may be, provided by law. Any and all moneys, and every valuable thing, or consideration of whatsoever kind, which shall be so as aforesaid, demanded, charged, bargained for, accepted, received or retained contrary to the provisions of this section, shall be deemed and held an additional and corrupt rate, charge or consideration for the water intended to be furnished or delivered therefor, or because thereof, and wholly extortionate and illegal, and when paid, delivered or surrendered, may be recovered back by the party paying, delivering or surrendering the same from the party to whom, or for whose use the same shall have been paid, delivered or surrendered, together with costs of suit, including reasonable fees of attorneys of plaintiff, by proper action in any court having jurisdiction. L. '87, p. 308, sec. 1.

1. This section, to and including sec. 2309, constitute what is known as the "anti-royalty act."

Section 2305. Penalty for collecting excessive rates. Every person owning or controlling, or claiming to own or control, any ditch, canal or reservoir, such as is mentioned in the first section of this act, who shall after demand in writing made upon him for the supply or delivery of water for irrigation, mining, milling or domestic purposes, to be delivered from the canal, ditch or reservoir owned, possessed or controlled by him, and after tender of the lawful rate of compensation therefor, in lawful money, demand, require, bargain for, accept, receive or retain from the party making such application any money or other thing of value, or any promise or contract, or any valuable consideration whatever, as such royalty, bonus, premium, prerequisite or condition precedent, as is by the provisions of this said first section prohibited, shall be deemed guilty of a misdemeanor, and, on conviction thereof, shall be punished by a fine of not less than one hundred dollars nor more than five thousand dollars, or imprisonment for a term not less than three months nor more than one year, or both such fine and imprisonment, in the discretion of the court. L. '87, p. 309, sec. 2.

1. Sec. 1 referred to in this section, is sec. 2304 hereof.

Section 2306. Refusal to deliver water—Penalty. Every person owning or controlling, or claiming to own or control any ditch, canal or reservoir, such as is mentioned in the first section of this act, who shall after demand in writing, made upon him for the supply or delivery of water for irrigation, mining, milling or domestic purposes, to be delivered from the canal, ditch or reservoir, owned, possessed or controlled by him, and after tender of the lawful rate of compensation therefor, in lawful money, refuse to furnish or carry and deliver from such ditch, canal or reservoir, any water so applied for, which water can or may be, by use of reasonable diligence in that behalf, and within the carrying or storage capacity of such ditch, canal or reservoir, be lawfully furnished and delivered, without infringement of prior rights, shall be deemed guilty of a misdemeanor, and upon conviction thereof, shall be punished by fine of not less than one hundred dollars, nor more than five thousand dollars, or imprisonment for a term of not less than three months, nor more than one year, or both such fine and imprisonment in the discretion of the court. L. '87, p. 309, sec. 3.

1. Section 1 referred to in this section is sec. 2304 hereof.

Section 2307. When corporation refuses to deliver water—Attorney general prosecute. When any corporation, in defiance or by attempted evasion of the provisions of this act, shall, after tender of the compensation hereinbefore provided **for, refuse** to deliver water, such as is mentioned in the third **section** of this act, to to any person lawfully entitled **to** apply therefor, it shall be the duty of the attorney general, upon request **of the** county commissioners of the proper county, or upon his otherwise receiving due notice thereof, to institute and prosecute to judgment and final determination proceedings in *quo warranto*, for the forfeiture **of the** corporate rights, privileges and franchises **of any** such corporation so offending, or by *mandamus* **or other** proper proceedings, to compel it **to its** duty **in that** behalf. **L. '87, p. 310, sec.** 4.

1. Section 3 referred to in this section is sec. 2306 hereof.

Section 2308. "Person" defined—Liability. The word "person" as used in this act shall include corporations and associations, and the plural as well as the singular number. And every officer **of a** corporation, or member of an association or co-ownership, and every **agent** violating any of the provisions of this act shall **be liable** to restore the unlawful consideration extorted and **be** punishable under the penal provisions of this **act,** the same as if the thing done in disobedience to **its** provisions were done for his own sole benefit and advantage. L. '87, **p.** 310, sec. 5.

Section 2309. Repeal. All laws and parts of laws in conflict **with** any of the provisions **of** this act are hereby **repealed. L. '87, p. 310,** sec. 6.

DIVISION IV.

WATER DISTRICTS.

Section 2310. Lands **watered constitute** districts. That the lands now irrigated, or **which may be** hereafter irrigated from ditches now taking water from the following described rivers or natural streams of the state of Colorado, are hereby declared **to** constitute irrigation districts. L. **'79,** p. 97, sec. 5; **G.** S. '83, p. 568, sec. 1741.

7—I. L.

1.   For water divisions, see sec. 2440.

Section 2311.   District No. 1.   That water district No. 1 shall consist of all lands in the state of Colorado irrigated by water taken from that portion of the South Platte river between the mouth of the Cache la Poudre river and the west boundary line of Washington county, and from the streams draining into the said portion of the South Platte river.   L. '79, p. 97, sec. 5, G. S. '83, p. 568, sec. 1742; amended L. '87, p. 303, sec. 1; amended L. '89, p. 212, sec. 13.

Section 2312.   District No. 2.   That District No. 2 shall consist of land irrigated from ditches taking water from the South Platte river and its tributaries, except Big Thompson, St. Vrain and Clear Creek, between the mouth of the Cache la Poudre and the mouth of Cherry creek.   L. '79, p. 97, sec. 7; G. S. '83, p. 568, sec. 1743.

Section 2313.   District No. 3.   That district No. 3 shall consist of all the lands irrigated from ditches taking water from the Cache la Poudre and it tributaries. L. '79, p. 98, sec. 8; G. S. '83, p. 568, sec. 1744.

Section 2314.   District No. 4.   That district No. 4 shall consist of lands irrigated from ditches taking water from the Big Thompson and its tributaries.   L. '79, p. 98, sec. 9; G. S. '83, p. 568, sec. 1745.

Section 2315.   District No. 5.   That district No. 5 shall consist of all lands irrigated from ditches taking water from the St. Vrain creek and its tributaries, except the Boulder, its tributaries and Coal creek.   L. '79, p. 98, sec. 10; G. S. '83, p. 569, sec. 1746.

Section 2316.   District No. 6.   That district No. 6 shall consist of all lands irrigated from ditches taking water from the Boulder and its tributaries and Coal creek.   L. '79, p. 98, sec. 11; G. S. '83, p. 569, sec. 1747.

Section 2317.   District No. 7.   That district No. 7 shall consist of all lands irrigated from ditches taking water from Clear creek and its tributaries.   L. '79, p. 98, sec. 12; G. S. '83, p. 569, sec. 1748.

Section 2318. **District No. 8.** That **district** No. 8 shall consist **of all lands irrigated by** ditches taking water from Cherry **creek, Plum creek and** Platte river and their tributaries, except Bear creek above district No. 2 and below the forks of **the** north and south branches of the South Platte river. L. '79, **p.** 98, sec. **13;** G. S. '83, p. 569, **sec. 1749.**

Section 2319. District No. 9. That district **No.** 9 shall consist of all lands irrigated by ditches taking water **from** Bear creek and its tributaries. **L. '79,** p. 98, sec. **14;** G. S. **'83,** p. 569, sec. 1750.

Section 2320. **District** No. 10—New districts **to** be formed by governor. That district No. 10 shall consist of all lands **irrigated** from ditches taking water from the Fountain **and its tributaries;** *Provided,* That said district shall not **extend beyond the** limits of El **Paso** county.

Other irrigation districts may be formed from time to time by the governor on petition of parties interested. **L.** '79, p. 98, sec. 15; G. S. '83, p. 569, **sec 1751.**

1. **Is** this section repealed by **L. '85, p. 256?** See title and secs. 1 and 2; also sec. 2444.

Section 2321. Introductory clause. That **the lands now, or** which may be hereafter irrigated from **ditches or canals,** taking **water from** any of **the following described** rivers, **or natural** streams in the **state of Colo-rado, are** hereby **declared to** constitute **irrigation districts.** L. '85, p. 256, **sec. 3.**

Section 2322. District **No. 11.** "Water **district** No. 11 shall consist of all lands irrigated by **water taken** from that portion of the Arkansas river above water **district No.** 12, and from the **streams** draining into the said portion of the Arkansas **river."** L. '85, **pp.** 256, 257, **sec. 4;** amended L. '89, **p. 370, sec.** 1.

Section 2323. District No. 12. That district No. **twelve (12) shall** consist of **all** lands irrigated from ditches, **or·canals** taking water from that part of the Arkansas river **lying** in·Fremont county; also all lands irrigated from ditches or canals taking water from the tributaries of said portion of the Arkansas river, except Grape creek and its tributaries. L. '85, p. 257, sec. 5.

Section 2324.   District No. 13.   That district No. thirteen (13) shall consist of all lands irrigated from ditches or canals taking water from Grape creek and its tributaries. L. '85, p. 257, sec. 6.

Section 2325.   **Water** district No. 14. "Water district No. 14 shall consist of all lands irrigated by water taken from that portion of the Arkansas river situated within the boundaries of Pueblo county, and from the streams draining into the said portion of the Arkansas river, except the St. Charles and Huerfano rivers and their tributaries, and except also that portion of the Fountain embraced in water district No. 10, and the streams draining into the said portion of the Fountain." L. '85, p. 257, sec. 7 ; amended L. '89, p. 370, sec. 2.

Section 2326.   District No. 15.   That district number fifteen (15) shall consist of all lands irrigated from ditches or canals taking water from the St. Charles and its tributaries. L. '85, p. 257, sec. 8.

Section 2327.   District No. 16.   That district number sixteen (16) shall consist of all lands irrigated from ditches or canals taking water from the Huerfano and its tributaries.   L. '85, p. 257, sec. 9.

Section 2328.   **Water** district No. 17.   Water district No. 17 shall consist of all lands irrigated by water taken from that portion of the Arkansas river below water district No. 14, and above the mouth of the Purgatoire river, and from the streams draining into the said portion of the Arkansas river, except the Apishapa river and its tributaries.   L. '85, p. 257, sec. 10; amended L. '89, p. 370, sec. 3.

Section 2329.   District No. 18.   That district number eighteen (18) shall consist of all lands irrigated from ditches and canals taking water from the Apishapa and its tributaries.   L. '85, p. 257, sec. 11.

Section 2330.   District No. 19.   That district number nineteen (19) shall consist of all lands irrigated

from ditches or canals taking water from the Purgatoire
and its tributaries. L. '85, p. 257, sec. 12.

Section 2331. Water district No. 20. Water district
number twenty shall consist of all lands irrigated by water
taken from that portion of the Rio Grande above the
mouth of the Rio Conejos, and from the streams draining into the said portion of the Rio Grande including
Piedra spring, Gato and San Francisco creeks, and all
other streams that would in time of flood flow into the
said portion of the Rio Grande, although at ordinary
stages the waters thereof might not flow upon the surface to the Rio Grande, except Alamosa river and its
tributaries and the La Jara and Trinchera creeks and
their tributaries; *Provided*, That nothing in this act
shall be construed as inconsistent with the provisions of
the acts creating water districts numbered twenty-five,
twenty-six and twenty-seven. L. '85, p. 258, secs. 13,
16; amended L. '87, p. 301, sec. 42; amended L. '89,
p. 218, sec. 1.

Section 2332. District No. 21. That district number twenty-one (21) shall consist of all lands irrigated
from ditches or canals taking water from the Alamosa
and La Jara creeks and their tributaries. L. '85, p. 258,
sec. 14.

Section 2333. District No. 22. That district number twenty-two (22) shall consist of all lands in the
state of Colorado irrigated from ditches or canals taking
water from Conejos creek and its tributaries. L. '85,
p. 258, sec. 15.

Section 2334. Water district No. 23. That water
district No. 23 shall consist of all lands in the state of
Colorado being, or to be irrigated from ditches or canals
taking water from the South Platte river, and from any
of its direct or indirect tributaries at any point or points
above water district No. 8 in the said state. L. '89, p.
212, sec. 9.

Section 2335. Water district No. 24. "Water
district No. 24 shall consist of all lands in the state of
Colorado irrigated by water taken from that portion of
the Rio Grande between the mouth of the Rio Conejos

and the Colorado state line, from the streams draining into the said portion of the Rio Grande and from Costilla creek and the streams draining into Costilla creek. L. '85, p. 258, sec. 17; amended L. '89, p. 370, sec. 4."

Section 2336. Water district No. 25. Water district No. 25 shall consist of all lands irrigated by water taken from the San Luis creek, Sand or Medano creek, Big Spring creek, Little Spring creek, North Zapato creek, South Zapato creek, Middle creek, Bear creek, Sierra Blanca creek, and all streams draining into said creeks, and all other streams between said Sand or Medano creek and the said Sierra Blanca creek. L. '85, p. 258, sec. 18; amended L. '89, p. 370, sec. 5.

Section 2337. District No. 26. That district No. twenty-six (26) shall consist of all lands irrigated from ditches or canals taking water from Saguache creek and its tributaries. L. '85, p. 258, sec. 19.

Section 2338. District No. 27. That district No. twenty-seven (27) shall consist of all lands irrigated from ditches or canals taking water from Tuttle, Carnero, La Garita and all other creeks and their tributaries which have their source of water supply in the La Garita mountains, and flow eastward into the San Luis valley. L. '85, p. 258, sec. 20.

Section 2339. District No. 28. That district No. twenty-eight (28) shall consist of all lands irrigated from ditches or canals taking water from the Tomichi and its tributaries. L. '85, p. 259, sec. 21.

Section 2340. District No. 29. That district No. twenty-nine (29) shall consist of all the lands lying in the state of Colorado irrigated from ditches or canals taking water from that part of the San Juan river and its tributaries which lie above the junction of the San Juan river and the Rio Piedra, and including the Rio Piedra. L. '85, p. 259, sec. 22.

Section 2341. District No. 30. That district No. thirty (30) shall consist of all lands lying in the state of Colorado irrigated from ditches or canals taking water

from that part of the Rio Las Animas river and its trib-
taries which lie in Colorado. L. '85, p. 259, sec. 23.

Section 2342. District No. 31. That district number thirty-one (31) shall consist of all lands in the state
of Colorado irrigated from ditches or canals taking water
from that part of the Los Pinos river and its tributaries
which lie in Colorado. L. '85, p. 259, sec. 24.

Section 2343. Water district No. 32. "Water district No. 32 shall consist of all lands in the state of
Colorado irrigated by water taken from those natural
streams which drain into the San Juan river, and are
not included in water districts numbers 29, 30, 31, 33
and 34." L. '85, p. 259, sec. 25; amended L. '89, p.
371, sec. 6.

Section 2344. District No. 33. That district number thirty-three (33) shall consist of all lands lying in
the state of Colorado irrigated from ditches or canals
taking water from the La Plata river and its tributaries
which lie in Colorado. L. '85, p. 259, sec. 26.

Section 2345. District No. 34. That district number thirty-four (34) shall consist of all lands lying in
the state of Colorado irrigated from ditches or canals
taking water from the Rio Mancos and its tributaries.
L. '85, p. 259, sec. 27.

Section 2346. District No. 35. That water district
number thirty-five (35) be and the same is hereby established, which water district shall consist of all lands lying in the county of Costilla, in this state, watered by
the Trinchera river and its tributaries. L. '87, p. 307,
sec. 1.

1. That all acts and parts of acts inconsistent with
the provisions of this act be and the same are hereby repealed. L. '87, p. 308, sec. 2.

Section 2347. Introductory clause. That the
lands now, or which may be hereafter, irrrigated from
ditches or canals taking water from any of the following
described rivers or natural streams in the state of Colorado are hereby declared to constitute irrigation districts.
L. '87, p. 313, sec. 2.

Section 2348. **District No. 36.** That district No. 36 shall consist of all the lands irrigated from water taken from the Blue river and its tributaries. L. '87, p. 313, sec. 3.

Section 2349. **District No. 37.** That district No. 37 shall consist of all lands lying in the State of Colorado irrigated by waters taken from the Eagle river and its tributaries. **L. '87, p. 313, sec. 4.**

Section 2350. District No. 38. That district No. 38 shall consist of all the lands lying in the state of Colorado irrigated by waters taken from the Roaring Fork river and its tributaries. L. '87, p. 313, sec. 5.

Section 2351. **District No. 39.** That district No. 39 shall consist of all the lands lying in the state of Colorado and located on the north side of the Grand river, and extending from the mouth of the Roaring Fork to the mouth of Rhone creek, all said land being irrigated by waters taken from the Grand river or its tributaries, viz: **Elk creek, Rifle creek** and Rhone creek. L. '87, p. 314, sec. 6.

Section 2352. **District No. 40.** That district number forty (40) shall consist of all lands irrigated from ditches or canals taking water from Crystal creek and **Smith's Fork** and their tributaries, and so much of all lands lying within the boundaries of Delta county as are irrigated from ditches or canals taking water from the Gunnison river and its tributaries, except lands irrigated from ditches and canals taking water from the Uncompahgre river. L. '87, p. 311, sec. 2.

Section 2353. District No. 41. That district number forty-one (41) shall consist of all lands irrigated from ditches or canals taking water from the Uncompahgre river and its tributaries, except so much as are within the boundary lines of Ouray county. L. '87, p. 311, sec. 3.

Section 2354. District No. 42. That district No. forty-two (42) shall consist of all lands irrigated from ditches or canals taking water from the Grand and Gunnison rivers and **their tributaries within the** county of Mesa. L. '87, p. **311, sec. 4.**

Section 2355. District No. 43. That water district number forty-three is hereby established, and shall consist of all lands irrigated by ditches taking **water from the** White river and its tributaries. L. 87, p. 307, **sec. 1.**

Section 2356. District No. 44. That water **district** No. 44 shall consist of all lands irrigated by water **taken** from that portion of the Yampa river above the **mouth** of the Little Snake river and below the mouth of **Forti**fication creek, and from **the streams draining into the** said portion of **the Yampa river.** L. '89, p. 211, sec. 2.

Section 2357. District No. 45. That water **district** No. 45 shall consist of **all lands situated on** the south side of the Grand **river and irrigated** from ditches or canals taking water **from the Grand river** and its tributaries, between the **mouth of Roaring Fork** river and the **north** line of Mesa county. L. '89, p. 213, sec. 17.

Section 2358. District No. 46. That water district No. 46 shall consist of all lands irrigated by water taken from **that** portion of the North Platte river above the mouth of Michigan creek, and from the streams draining into said portion of the North Platte river. L. '89, p. 212, sec. 10.

Section 2359. District No. 47. **That** water district No. 47 shall consist **of all** lands in the state of Colorado irrigated by water **taken** from that portion of the North Platte river between **water** district No. 46 and the **state** line of Colorado, **and from** the streams draining into **the** said portion of the **North** Platte river, and from **Granite** and Encampment **creeks** and the streams draining **into** the said creeks. L. '89, p. 212, sec. **11.**

Section 2360. **District No.** 48. That water **dis**trict No. 48 shall consist **of all** lands in the state of **Colorado** irrigated by water taken from the Big Laramie river and from the streams draining into the said river. L. 89, p. 212, sec. 12.

Section 2361. District No. 49. That water district No. 49 shall **consist** of all **lands** in the state of Colorado

irrigated by water taken from the south fork of the
Republican river and the Smoky Hill river, and the
streams draining into said rivers. L. '89, p. 471, sec. 1.

Section 2362. District No. 50. That water district
No. 50 shall consist of all lands irrigated by water taken
from the Muddy and Troublesome creeks, and from the
streams draining into said creeks. L. '89, p. 213,
sec. 18.

Section. 2363. District No. 51. That water district
No. 51 shall consist of all lands irrigated by water taken
from the Grand river above the mouth of the Blue river,
and from the streams draining into the said portion of
the Grand river, except the Muddy and Troublesome
creeks, and the streams draining into said creeks. L.
'89, p. 213, sec. 19.

Section 2364. District No. 52. That water district No. 52 shall consist of all lands on the south side
of the Grand river irrigated by water taken from the
Grand river below the mouth of the Blue river and
above the mouth of Roaring Fork river, and from the
streams draining into the said portion of the Grand
river, except Eagle river and its tributaries. L. '89, p.
213, sec. 20.

Section 2365. District No. 53. That water district No. 53 shall consist of all lands on the north
side of Grand river irrigated by water from that portion
of the Grand river below the mouth of Muddy creek
and above the mouth of Roaring Fork river, and from
the streams draining into the said portion of the Grand
river. L. '89, p. 214, sec. 21.

Section 2366. District No. 54. That water district No. 54 shall consist of all lands in the state of Colorado irrigated by water taken from that portion of the
Little Snake river and its tributaries above the most
westerly intersection of said river with the Colorado
state line. L. '89, p. 211, sec. 3.

Section 2367. District No. 55. That water district No. 55 shall consist of all lands in the state of Col-

orado irrigated by water taken from that portion of the Yampa river below water district No. 44, and from the streams draining into the said portion of Yampa river not included in water district No. 54. L. '89, p. 211, sec. 4.

Section 2368. District No. 56. That water district No. 56 shall consist of all lands in the state of Colorado irrigated by water taken from that portion of the Green river embraced within the boundaries of the county of Routt, and from the streams draining into the said portion of the Green river, except the Yampa river and its tributaries. L. '89, p. 211, sec. 5.

Section 2369. District No. 57. That water district No. 57 shall consist of all lands irrigated by water taken from that portion of the Yampa river above water district No. 44 and below the mouth of Elk creek, and from the streams draining into the said portion of the Yampa river. L. '89, p. 211, sec. 6.

Section 2370. District No. 58. That water district No. 58 shall consist of all lands irrigated by water taken from the Yampa river above water district No. 57 and from the streams draining into the said portion of Yampa river. L. '89, p. 211, sec. 7.

Section 2371. District No. 59. That water district No. 59 shall consist of all lands irrigated by water taken from the Gunnison river above the mouth of the Tomichi creek; and from all streams draining into the said portion of Gunnison river; also of all lands on the north side of Gunnison river irrigated by water taken from the Gunnison river below the mouth of Tomichi creek and above water district No. 40, and from the streams draining into the said portion of the Gunnison river. L. '89, p. 214, sec. 22.

Section 2372. District No. 60. That water district No. 60 shall consist of all lands irrigated by water taken from the San Miguel river and from the streams draining into the said river. L. '89, p. 214, sec. 23.

Section 2373. District No. 61. That water district No. 61 shall consist of all lands in the state of

Colorado irrigated by water taken from that portion of the Dolores river above the mouth of San Miguel river and from the streams draining into the said portion of the Dolores river. L. '89, p. 214, sec. 24.

Section 2374. District No. 62. That water district No. 62 shall consist of all lands south of the Gunnison river irrigated by water taken from the Gunnison river below the mouth of Tomichi creek and above water district No. 40, and from the streams draining into the said portion of the Gunnison river. L. '89, p. 214, sec. 25.

Section 2375. District No. 63. That water district No. 63 shall consist of all lands in the state of Colorado irrigated by water taken from that portion of the Dolores river below the mouth of the San Miguel river and from the streams draining into the said portion of the Dolores river. L. '89, p. 214, sec. 26.

Section 2376. District No. 64. That water district No. 64 shall consist of all lands irrigated by water taken from that portion of the South Platte river between the western boundary line of Washington county and the state line of Colorado and Nebraska, and from the streams draining into the said portion of the South Platte river. L. '89, p. 213, sec. 14.

Section 2377. District No. 65. That water district No. 65 shall consist of all lands in the state of Colorado irrigated by water taken from the middle and north forks of the Republican river, from Sandy and French-man's creeks, and the tributaries of these streams. L. '89, p. 213, sec. 15.

Section 2378. District No. 66. That water district No. 66 shall consist of all lands in the state of Colorado irrigated by water taken from the Dry Cimmarron and the streams draining into the said river. L. '89, p. 472, sec. 2.

Section 2379. District No. 67. That water district No. 67 shall consist of all the lands in the state of Colorado irrigated by water taken from that portion of the Arkansas river below the mouth of the Purgatoire river, and from the streams draining into the said portion of the Arkansas river. L. '89, p. 472, sec. 3.

Section 2380. District No. 68. That water district No. 68 shall consist of all lands irrigated by water taken from that portion of the Uncompahgre river above water district No. 41, and from the streams draining into the said portion of the Uncompahgre river. L. '89, p. 213, sec. 16.

DIVISION V.

WATER COMMISSIONERS.

Section 2381. Number of water commissioners—How appointed—Bonds—Term of office. There shall be one water commissioner for each of the above named districts and for each district hereafter formed, who shall be appointed by the governor, to be selected by him from persons recommended to him by the several boards of county commissioners of the counties into which water districts may extend; and the water commissioner so appointed shall, before entering upon his duties, give a good and sufficient bond for the faithful discharge of his duties, with not less than three sureties, in a sum not less than one thousand nor more than five thousand dollars, the amount of said bond to be fixed by the county commissioners and approved by the governor and state engineer. The commissioner so appointed shall hold his office until his successor is appointed and qualified; *Provided, however,* That if such water district shall be embraced in more than one county, and the several counties in which such water district is situated disagree as to the amount of the bond as herein required of water commissioners, then and in that event the governor shall fix the amount thereof, with the same effect as though fixed by the county commissioners. L. '79, pp. 98, 99, sec. 16; G. S. '83, p. 569, sec. 1752; amended L. '87, p. 302, sec. 1.

2. The division made by water commissioners is subject to review by the courts. *Certiorari* lies if the commissioners exceed their jurisdiction. See Code '87, sec. 297.

Section 2382. **Vacancies—How** filled—Removal. The governor shall, by like **selection** and appointment, fill all vacancies which **may be** occasioned by death, resignation or continued **absence** from the district, removal or otherwise. Said county commissioners may, from time to time, recommend persons to be appointed as above provided, and the governor may at any time remove any water commissioner in his discretion. L. '87, p. 302, sec. 2.

Section 2383. Oath of office within ten days. That within ten days after his appointment and before entering upon the duties of his office such water commissioner shall take and subscribe the oath of office prescribed by the constitution of this state. L. '79, p. 99, sec. 17; G. S. '83, p. 570, sec. 1753.

Section 2384. Duty of water commissioners—Open and shut headgates. It shall be the duty of said water commissioners to divide the **water in the natural** stream or streams of their district **among the several** ditches taking water **from** the **same**, according to the prior rights of **each respectively, in** whole or in part to shut and fasten, **or** cause to be **shut** and fastened by order given to any sworn assistant, sheriff or constable of the county in which the head of such ditch is situated, the headgates of any ditch or ditches heading in any of the natural streams of the district, which in a time of scarcity of water, shall **not** be entitled **to** water by reason **of the** priority of **the rights** of others below them on the same stream. L. '79, p. 99, sec. 18; G. S. '83, p. 570, sec. 1754.

Section 2385. **Interfering** with headgate or water box—Penalty. Every person who shall wilfully open, **close,** change **or** interfere with any headgate or water box without authority, shall be guilty of a misdemeanor, and on conviction thereof, shall be fined not less **than** fifty dollars nor **more than** three hundred dol-

lars, and may be **imprisoned not less than sixty days.**
L. '79, p. 108, sec. 44; **G. S. '83, p. 570, sec.** 1755.

1.   Water commissioners **invested with police pow-**
ers.   Sec. 2386, sec. 2391.

Section 2386.   Power of water commissioners —
State engineer.   Water commissioners shall, in the discharge of their duties, be invested with the powers **of**
constables, and may arrest any person violating his
(their) orders relative to the opening or shutting down
of headgates, or **the** using of water for irrigating purposes, and take **such** offender before the **nearest** justice
of **the** peace, who **may, if such** offender **be convicted,**
fine him in any sum **not exceeding** one hundred dollars,
and, in default of **the** payment **of** such fine, may im-
prison him in the county jail **not** exceeding thirty days;
Provided, That the orders of the superintendents of irri-
gation in their respective divisions, and the orders of the
state engineer, shall be **held** at all times superior to the
orders of water commissioners, and shall relieve any per-
son acting in accordance with **such** superior orders from
the penalties herein provided; *And provided,* Also, that
in like manner the orders issued **by** the state engineer
shall be held superior to any order issued by any **super-**
intendent of irrigation.   L. '89, p. 469, sec. 1.

Section 2387.   **Pay of** water commissioners—Veri-
fied account.   The **water** commissioners shall be entitled
to pay at the rate **of five** (5) dollars per day for each day
he shall actually **be** employed in the duties of his office,
and be paid by the county or counties in which his irrigating district may lie.   Each water commissioner shall
keep a just and itemized account of the time spent by
him **in** the duties of his office, and shall present a true
copy thereof, verified by oath, to the board of county
commissioners of the county in which his district may
lie, and said board of commissioners shall allow the
same; and if said irrigation district shall extend into
two or more counties, then such water commissioner
shall present his account for his services, verified as
aforesaid, to **the** board of county commissioners into
which his district extends, **and** each board of county

commissioners shall pay its *pro rata* share thereof.   L.
'89, p. 470, sec. 2.

Section 2388.   Employ suitable assistance—Pay.
The water commissioner is hereby given power, whenever he shall deem it necessary, to employ a suitable
assistant or assistants, to aid him in the discharge of his
duties; such assistant or assistants shall take the same
oath as water commissioner, and shall obey his instruct-
tions, and shall be entitled to pay at the rate of two
dollars as (and) fifty cents ($2.50) per day for every day
they are so employed, to be paid by county commissioners upon the certificates of the water commissioners.
L. '89, p. 470, sec. 3.

Section 2389.   Itemized account of time.   Each
water commissioner shall keep an itemized account of
the time of each assistant by him employed, and shall
certify the same to the board of county commissioners,
who shall pay such assistant or assistants, in the same
manner as provided for payment of water commissioners in section two of this act.   L. '89, p. 470, sec. 4.

1.   Section 2 referred to is sec. 2387 hereof.

Section 2390.   Repeal.   That section one of an
act entitled, "An act to amend and (an) act entitled an
act to regulate the use of water for irrigation, and providing for settling the priority of rights thereto, and for
payment of the expenses thereof, and for payment of all
costs and expenses incident to said regulations of use,"
approved February 19, 1879; approved April 9, 1885;
and also sections forty-one of an act entitled, "An act to
regulate the use of water for irrigation and providing
for settling the priority of rights thereto, and for payment ot the expenses thereof, and for payment of all
costs and expenses incident to said regulation of use,"
approved February 19, 1879, and all other acts inconsistent herewith, are hereby repealed.   L. '89, p. 470,
sec. 5.

Section 2391.   Commissioner devote entire time to
duties—Penalty for neglect.   It is hereby made the duty
of the water commissioner, after being called upon to
distribute water, to devote his entire time to the dis-

charge of his **duties when such duties** are required, so long as the necessities **of irrigation in his** district shall require; and it is made **his duty to be** actively employed on the line of the stream or streams in his water district, supervising and directing the putting in of headgates, wastegates, keeping the stream clear of unnecessary **dams or other** obstructions, and such other duties **as pertain to a** guard of the public streams in **his water district;** and for wilful neglect of his duty, he **shall be liable to** fifty dollars fine, with costs of suit. **L. 89, p. 471, sec. 6.**

Section 2392. **Commissioner** begin **work when called on.** Said water **commissioners** shall **not begin their work** until they **shall be** called on by two **or more owners or managers,** or persons controlling ditches in their several districts, by application in writing, **stating that** there **is necessity** for their **action;** and they **shall not** continue performing services after the **necessity** therefore (therefor) shall cease. L. '79. pp. 107, **108, sec.** 43; G. S. '83, pp. 570, 571, sec. 1758.

1. Commissioner failing **to perform** duty, **sec. 2291.**

DIVISION VI.

OFFENSES.

Section 2393. Cutting or breaking gate, **bank, side of ditch,** flume, etc.—Penalty. Any person or **persons who** shall knowingly and wilfully **cut,** dig, break **down** or open **any gate,** bank, embankment or side **of** any ditch, **canal, flume,** feeder **or reservoir** in **which** such person **or persons may be a joint** owner, **or** the property of **another, or in the lawful** possession of another or others, and used **for the** purpose of irrigation, manufacturing, mining or domestic purposes, with intent maliciously to injure **any** person, association or **corporation,** or for his or her own gain, unlawfully, with **intent of stealing,** taking or causing to run or pour out **of such** ditch, canal, reservoir, feeder or flume, any **water for his or her** own profit, benefit or advantage, to the injury of any other person, persons, association or corporation, lawfully in the use of such water or of such ditch, canal, reservoir, feeder or flume, he, she or they so offending shall be deemed guilty of a misdemeanor, and on conviction thereof shall be fined in any sum not less than five dollars nor more than three hundred dollars, and may be imprisoned in the county jail not exceeding ninety days.   L. '81, p. 163, sec. 1; G. S. '83, p. 571, sec. 1759.

1.   Penalty for polluting water in ditch or stream. Secs. 1357, 3960, 3962.

2.   Penalty for damaging any ditch, flume, etc. Sec. 574.

3.   Bribing person in charge of distribution of water.   Sec. 2398.

4.   As to right to take water without knowledge of ditch company.   See Coffin v. Left-Hand Ditch Co., 6 Colo., 444-5 (1882).

5.   A person holding an assignment of shares of stock in a joint-stock ditch company, but not transferred on the books of the company, is not entitled to water from a ditch for the irrigation of his lands, not having used water therefrom; and if he takes water by force from the ditch he is liable in trespass.   Supply Ditch Co. v. Elliott, 10 Colo., 330-5 (1887).

Section 2394.   Jurisdiction of justices of the peace. Justices of the peace shall have the jurisdiction of all offenses under the provisions of this act, saving to any party defendant the right to be tried by a jury, as in other criminal cases before such justices now provided for by law; and also the right to appeal in manner and form as by law, now or hereafter provided by law, in criminal cases before such justices.   L. '81, p. 163, sec. 2; G. S. '83, p. 571, sec. 1760.

Section 2395.   No person to receive more water than he is entitled to.   That it shall be the duty of every person who is entitled to take water for irrigation purposes from any ditch, canal or reservoir to see that he receives no more water from such ditch, canal or reservoir through his headgate, or by any ways or means whatsoever, than he is entitled to, and that he shall, at all times, take every precaution to prevent more water than he is entitled to coming from such ditch, canal or reservoir upon his land.   L. '87, p. 312, sec. 1.

Section 2396.   Duties of parties taking water—Liability—Damages—Costs.   That it shall be the duty of every such person taking water from any ditch, canal or reservoir, to be used for irrigating purposes, on finding that he is receiving more water from such ditch, canal or reservoir, either through his headgate or by means of leaks, or by any means whatsoever, immediately to take steps to prevent his further receiving more water from such ditch, canal or reservoir than he is entitled to, and if knowingly he permits such extra water to come upon his land from such ditch, canal or reservoir, and does not immediately notify the owner or owners of such ditch, or take steps to prevent its further flowing upon his land, he shall be liable to any person, company, or corporation who may be injured by such extra appropriation of water for the actual damage sustained by the party aggrieved, which damages shall be adjudged to be paid, together with the costs of suit and a reasonable attorney's fee, to be fixed by the court and taxed with the costs.   L. '87, p. 312, sec. 2.

Section 2397.   Ditches free from taxation.   That all ditches used for the purpose of irrigation, and that only where the water is not sold for the purpose of deriving a revenue therefrom, be and the same are hereby declared free from all taxation, whether for state, county or municipal purposes.   L. '72, p. 143, sec. 1; G. L. '77, p. 517, sec. 1384; G. S. '83, p. 571, sec. 1761.

1.   See sec. 3766, chap. "Revenue;" also const., art. X., sec. 3.

2.   This section has no place under this division, but it is left there as found in G. S. '83, p. 571.

Section 2398.   Dishonest distribution of water—A misdemeanor—Penalty.   Any water commissioner, or any deputy water commissioner, assistant, water master, superintendent, ditch **rider or** other person in charge of the divisions or distributions of water, whether from the public stream or from any ditch or canal, who shall take or receive any money, promises or favors, or anything of value, intended to influence him dishonestly to favor, or cause water to accrue or run to any person or persons' advantage, benefit or gain, detrimental to the rights of others, shall be deemed guilty of a misdemeanor, and shall be fined in any sum not less than fifty (50) dollars nor more than three hundred (300) dollars. Any person giving or offering any such money, promises or favors, or any other thing of value, to any of such above-named persons, with intent as aforesaid, shall likewise be deemed guilty of a misdemeanor, and upon conviction thereof shall be punished by a fine in any sum not less than fifty (50) dollars nor more than three hundred (300) dollars; and any fines so collected shall be paid into the school fund of the county wherein such fines are collected. L. '89, p. 39, sec. 1.

DIVISION VII.

ADJUDICATION OF RIGHTS.

Section 2399. Jurisdiction of courts—How vested. For the purpose of hearing, adjudicating and settling all questions concerning the priority of appropriation of water between ditch companies and other owners of ditches drawing water for irrigation purposes from the same stream or its tributaries within the same water district, and all other questions of law and questions of right growing out of or in any way involved or connected therewith, jurisdiction is hereby vested exclusively in the district court of the proper county; but when any water district shall extend into two or more counties, the district court of the county in which the first regular term after the first day of December in each year shall soonest occur, according to the law then in force, shall be the proper court in which the proceedings for said purpose, as hereinafter provided for, shall be commenced; but where said proceedings shall be once commenced, by the entry of an order appointing a referee in the manner and for the purpose hereinafter in this act provided, such court shall thereafter retain exclusive jurisdiction of the whole subject until final adjudication thereof is had, notwithstanding any law to the contrary now in force. L. '79, pp. 99, 100, sec. 19; G. S. '83, pp. 571, 572, sec. 1762.

1. An act entitled "an act to regulate the use of water for irrigation, and providing for settling the priority of right thereto, and for payment of the expenses thereof, and for payment of all costs and expenses incident to said regulation of use;" approved and in force February 19, 1879 (L. '79, p. 94, etc.), was never specifically repealed. But many sections of it are supplanted by the act at the head of this division on VII. (L. '81, p. 142); but where there is no necessary conflict between the two acts, the sections of both are to be found in our text.

2. The title of the said act (L. '79, p. 94) set out in the foregoing note, is sufficient and clearly expresses the subject as required by Const., art. V., sec. 21; Golden Canal Co. v. Bright, 8 Colo., 147 (1884).

3. The irrigation acts of 1879 and 1881 were intended as a system of procedure for determining the priority of rights to the use of water for irrigation between the owners of the ditches, canals and reservoirs taking water from the same natural stream. Platte W. Co. v. North Colo. I. Co., 12 Colo., 529 (1889). See Mills' Const., Ann., art. XVI., sec. 6, note 48, etc.

4. As to the divided position of our supreme court on the purpose of this statute adjudicating priorities, whether it is simply a police regulation or not, see Southworth case, 13 Colo., 135, etc. (1889), in notes 93-6 of Const., art. XVI., sec. 6.

Section 2400. Filing statements of claims—Ditch, name, description, post-office address. In order that all parties may be protected in their lawful rights to the use of water for irrigation, every person, association or corporation owning or claiming any interest in any ditch, canal or reservoir, within any water district, shall, on or before the first day of June, A. D. 1881, file with the clerk of the district court having jurisdiction of priority of right to the use of water for irrigation in such water district, a statement of claim, under oath, entitled of the proper court, and in the matter of priorities of water rights in district No.____, as the case may be, which statement shall contain the name or names, together with the post-office address of the claimant or claimants claiming ownership, as aforesaid, of any such ditch, canal or reservoir, the name thereof (if) any, and if without a name, the owner or owners shall choose and adopt a name, to be therein stated, by which such ditch, canal or reservoir shall thereafter be known, the description of such ditch, canal or reservoir, as to location of headgate, general course of ditch, the name of the natural stream from which such ditch, canal or reservoir draws its supply of water, the length, width, depth and grade thereof, as near as may be; the time, fixing a day, month and year as the date of the appropriation of water by original construction, also by any enlargement or extension, if any such thereof may have been made, and the amount of water claimed by or under such construction, enlargement or extension, and the present capacity of the ditch, canal or feeder of reservoir, and also the number of acres of land lying under and being or proposed to be irrigated by water from such ditch, canal or reservoir. Said statement shall be signed by the proper party or parties.   L. '81, pp. 142, 143, sec. 1; G. S. '83, p. 572, sec. 1763.

1. Referee's notice to file statement, sec. 2410; clerk publishes notice, sec. 2405; see sec. 2415; see also sec. 2265.   This section referred to in F. H. L. C. & R. Co. v. Southworth, 13 Colo., 134 (1879).   For digest of the case see Mills' Const. Ann., art. XVI, sec. 6, note 93, etc.

Section 2401.   Secretary of state make publication —Publisher's certificate.   The secretary of state shall, without delay, after the passage of this act, cause a certified copy of the foregoing section, giving the date of the approval of this act, to be published in one of the public newspapers published in such county in which part or portion of any water district is or shall be established by law at the time of such publication; and said sec. 1 shall be published, as aforesaid, once in each and every week continuously in said paper until said first day of June, A. D. 1881, and in case in the meantime any one of said papers shall cease to be published, then such publication shall be made in some other paper in the same county (if any); and on conclusion of such publication such publisher of such paper shall deliver to the secretary of state his sworn certificate of publication in duplicate, showing that such publication has been made in his paper in compliance with the preceding section hereof, and stating the first and last day of such publication; and he shall thereupon be entitled to receive from the secretary of state a certificate of the amount due him for such publication, on presentation of which to the auditor of state he shall draw his warrant for the amount in favor of the holder on the state treasurer, who shall pay the same according to law. L. '81, pp. 143, 144, sec. 2; G. S. '83, pp. 572, 573, sec. 1764.

Section 2402. Secretary's certificate—Where filed —Effect. The secretary of state shall file one of said duplicate certificates of publication with the clerk of the district court having jurisdiction of priority of rights to use of water for irrigation in the proper water district, certifying officially that such publication therein mentioned was duly authorized by him, and said clerk shall file the same with the statement of claim provided for in one section hereof, and such certificate of such publisher or any additional certificate of the same publisher to same fact, in case of loss of the original, shall be proof of the proper publication of said section in the paper therein mentioned. Said secretary of state shall also certify to such clerk of the several district courts having jurisdiction of said priorities of right to use of water for irrigation throughout the state the names of the newspapers and of the county in which he caused such publication to be made, and that the duplicate certificate of publication of the publisher, as herein required, are (is) on file in his office, and said certificate shall be sufficient proof of the publication of said sec. 1 hereof, as by this act required.   L. '81, p. 144, sec. 3; G. S. '83, p. 573, sec. 1765.

1.   Sec. 1, referred to, is sec. 2400 hereof.

Section 2403.   Proceedings in court—Order—Evidence — Examination — Proofs—What facts—Decree—Certificate of clerk.   When, at any time after the first day of June, A. D. 1881, any one or more persons, associations or corporations, interested as owners of any ditch, canal or reservoir in any water district, shall present to the district court of any county having jurisdiction of priority of rights to the use of water for irrigation in such water district, according to the provisions of an act entitled an act to regulate the use of water for irrigation and providing for settling the priority of rights thereto, and for payment of the expenses thereof, and for payment of all costs and expenses incident to said regulation of use, or to the judge thereof in vacation, a motion, petition or application in writing, moving or praying said court to proceed to an adjudication of the priorities of rights to use of water for irrigation between the several ditches, canals and reservoirs in such district, the court or judge thereof in vacation, shall, without unnecessary delay, in case he shall deem it practicable to proceed in open court, as prayed for, by an order to be entered of record upon such motion, petition or application, appoint a day in some regular or special term of said court, for commencing to hear and take evidence in such adjudication, at which time it shall be the duty of the court to proceed to hear all evidence which may be offered by or on behalf of any person, association or corporation interested in any ditch, canal or reservoir in such district, either as owner or consumer of water therefrom, in support of or against any claim or claims of priority of appropriation of water made by means of any ditch, canal or reservoir, or by any enlargement or extension thereof, in such district, and consider all such evidence, together with any and all evidence, if any, which may have been heretofore offered and taken in such district in the same matter by any referee heretofore appointed under the provisions of said act above herein mentioned, and also the arguments of parties or their counsel, and shall ascertain and find from such evidence, as near as may be, the date of the commencement of such ditch, canal or reservoir, together with the original size and carrying capacity thereof as originally constructed, the time of the commencement of each enlargement or extension thereof, if any, with the increased capacity thereby occasioned, the time spent, severally, in such construction and enlargement, or extension and re-enlargement, if any, the diligence with which work was in each case prosecuted, the nature of the work as to difficulty of construction, and all such other facts as may tend to show the compliance with the law, in acquiring the priority of right claimed for each such ditch, canal or reservoir, and determine the matters put in evidence, and make and cause to be entered a decree determining and establishing the several priorities of right, by appropriation of water, of the several ditches, canals and reservoir (reservoirs) in such water district, concerning which testimony shall have been offered, each according to the time of its said construction and enlargement, or enlargements or extensions, with the amount of water which shall be held to have been appropriated by such construction and enlargements, or extensions, describing such amount by cubic feet per second of time, if the evidence shall show sufficient data to ascertain such cubic feet, and, if not, by width, depth and grade, and such other description as will most certainly and conveniently show the amount of water intended as the capacity of such ditch, canal or reservoir in such decree. Said court shall further order that each and every party interested or claiming any such ditch, canal or reservoir shall receive from the clerk, on payment of a reasonable fee therefor, to be fixed by the court, a certificate, under seal of the court, showing the date or dates and amount or amounts of appropriations adjudged in favor of such ditch, canal or reservoir, under and by virtue of the construction, extension and enlargements thereof, severally; also specifying the number of said ditch and of each priority to which the same may be entitled by reason of such construction, extension and enlargements. L. '81, pp. 144-146, sec. 4 ; G. S. '83, pp. 573, 574, sec. 1766.

1. The irrigation acts of 1879 and 1881 were intended as a system of procedure for determining the priority of rights to the use of water for irrigation between the owners of ditches, canals and reservoirs taking water from the same natural stream. Platte W. Co. v. North. Colo. Irr. Co., 12 Colo., 529 (1889); see Const., art. XVI., sec. 6, note 48, etc.

2. This section referred to in F. H. L. & C. & R. Co. v. Southworth, 13 Colo., 135 (1889).

3. For digest of this case, see Mills' Const. Ann., art. XVI., sec. 6, note 93, etc.

Section 2404. Copy of decree—Authority of commissioner—Recording—Copy—Evidence. The holder of such certificate shall exhibit the same to the water commissioner of the district when he commences the exercise of his duties, and such water commissioner shall keep a book in which shall be entered a brief statement of the contents of such certificate, and which shall be delivered to his successor, and said certificate, or statement thereof in his book, shall be the warrant of authority to said water commissioner for regulating the flow of water in relation to such ditch, canal or reservoir. Said certificate shall be recorded at the same rates of charges as in cases of deeds of conveyance, in the records of each county into which the ditch, canal or reservoir, to which such certificate relates, shall extend; and said certificate, or said record thereof, or a duly certified copy of such record, shall be *prima facie* evidence of so much of said decree as shall be recited therein, in any suit or proceeding in which the same may be relevent. L. '81, pp. 146, 147, sec. 5; G. S. '83, pp. 574, 575, sec. 1767.

Section 2405. Clerk publish notice—Post copy— Ten copies posted by party petitioning. Notice shall be given by the clerk of said court, of the time so appointed, by publishing the same in one public newspaper in such county into which such water district may extend, which notice shall be so published in such paper once in each **week until four successive** weekly publications shall **have been made, the last of** which shall be on a day previous **to** the **day appointed as afore-said.** Said notice shall contain a **copy of said order,** and shall notify **all** persons, associations **and corporations** interested as owners in any ditch, **canal or reservoir in such** water district, to appear at said **court at the time** so appointed and file a statement of claim, under **oath,** in case no statement has been before filed by him, **her** or them, showing the ditch, canal or reservoir, or **two or more** such, in which **he,** she or they claim an interest, together with **the names of all** the owners thereof, which statement **may be made by any one** of **the own-**ers of such ditch, **canal or reservoir, for and in behalf of** all; and also that **all persons interested as** owners **or** consumers may then **and** there present his, **her or** their proofs for or against **any** priority of right **of** water **by** appropriation sought **to** be shown by any party by **or** through any such ditch, canal **or** reservoir (either as **owner or** consumer of water drawn therefrom). Ten printed copies of said notice shall be posted in ten pub-lic places in such water district, **not** less than twenty days before the day so appointed, **which** copies shall be **so** posted by the party or parties moving the adjudica-cation. L. '81, p. **147,** sec. 6; G. S. '83, p. **575,** sec. **1768;** see sec. 2410.

Section 2406. **Proof** of publication—Of posting copies—Entry by **clerk.** Proof of the proper publica-tion **of** said notice **or** notices in said public papers shall consist in such case **of the sworn** certificate of the pub-lisher **of** such newspaper, **showing** the publication to have been made **in** accordance **with** the provisions of section three of **this** act, which **certificate** shall be pro-cured by the party **or** parties **moving the** adjudication, **at his** or their expense, and **on said certificate** being **filed** with the **clerk,** shall enter **the** amount of the printer's fee therefor as costs advanced by the party procuring the same, which sum shall be counted to his, her **or** their credit in distribution **of** costs. Proof of **the posting of** said printed copies shall be made by the **affidavit of** some credible person, certified to be such by **the clerk or other** officer administering the oath, show-

ing when, where and how said copies were posted. L.
'81, pp. 147, 148, sec. 7; G. S. '83, p. 575, sec. 1769.

1.   Section 3 referred to is sec. 2402 hereof.

Section 2407.   Notices served on all parties—How
served—Notice by mail.   The party or parties moving
such adjudication shall cause a printed or written copy
of the notice aforesaid, published as aforesaid, to be
served on every person, association or corporation shown
by the statement of claim on file, as provided in section
1 hereof; which service shall be made within ten days
from the time of the first publication by the clerk, by
any credible person certified by said clerk or referee to
be such, by delivering such copy as aforesaid to the person to be served, if such person, by due diligence can
be found in the county of his residence.   If such person
can not be found, as aforesaid, then by leaving such
copy at his or her usual place of residence, if he or she
have such residence, in charge of some person of the
age of fourteen years or over, there residing; and on any
corporation, by delivering the copy to the president or
vice-president, or secretary or treasurer thereof, or the
manager or superintendent in charge of their ditch,
canal or reservoir, or authorized agent or attorney, or
by leaving such copy at the office or usual place of business of such corporation; and the proof of such service
shall be made by affidavit of the person or persons
serving said copies, showing when and how such service
has been made on such party.   In case of parties not
served in any manner as aforesaid, the clerk shall
deposit in the postoffice, duly enclosed in an envelope,
with the proper postage stamp thereon, a copy directed
to the address of such party, shown in the statement of
claim aforesaid, filed by him or her under section one
hereof.   L. '81, pp. 148-149, sec. 8; G. S. '83, pp. 575-
576, sec. 1770.

1.   Section 1 referred to is sec. 2400 hereof.

Section 2408.   Decrée—Court number all ditches—
Reservoirs—Number appropriations.   The court, in
making such decree, as aforesaid, shall number the several ditches and canals in the water district, concerning

which adjudication is made, in consecutive order, according to the priority of appropriation of water thereby made by the original construction thereof, as near as may be, having reference to the date of each decree as rendered, and also number the reservoirs in like manner separately from ditches and canals, and shall further number each several appropriation of water consecutively, beginning with the oldest appropriation, without respect to the ditches or reservoirs by means of which such appropriations were made ; whether such appropriation shall have been made by means of construction, extension or enlargement, which of each ditch, canal or reservoir, together with the number or numbers of any appropriations of water held to have been made by means of construction, extension or enlargement thereof, shall be incorporated in said decree and certificate of the clerk, to be issued to the claimants, as provided in section one of this act, so as to show the order in priority of such ditch or canal, and of such reservoir, and also of such successive appropriation of water pertaining thereto, for the information of the water commissioner of the district in distributing water ; such numbering to be as near as may be having reference to date of decrees as rendered. L. '81, p. 149, sec. 9; G. S. '83, p. 576, sec. 1771.

1. Section 1 referred to is sec. 2400 hereof.

DIVISION VIII.

REFEREE.

Section 2409. When court may appoint referee— What referred. If for any cause the judge of said court shall deem it impracticable or inexpedient to proceed to hear such evidence in open court, he shall, instead of the order mentioned in section four of this act, make and cause to be entered of record an order appointing some discreet person properly qualified, a referee of said court, to whom shall be referred the statement of claim aforesaid on file in said matter, the matter of taking evidence and reporting the same, making an abstract and findings upon the same, and preparing a decree in said adjudication; and also in the case of any water district in which a referee has been heretofore appointed, and evidence taken by him under the provisions of this act, the title of which is recited in section four of this act; such evidence so already taken, together with the abstract thereof, and report to the referee who took the same, shall be also referred to said referee, to be appointed as aforesaid, and he shall proceed with his duties as hereinafter provided, first taking an oath (of) office, such as is required to be taken by referees in other cases under the provisions of the code of civil procedure. L. '81, pp. 149, 150, sec. 10; G. S. '83, pp. 576, 577, sec. 1772.

1. Section 4 referred in this section is sec. 2403 hereof.

2. Where a referee was appointed to take testimony by a district judge, and the rules made by said judge in relation to the taking of proofs to adjudicate priorities are alleged to be inadequate, illegal, etc., mandamus will not lie to compel said judge to change said rules. Union Colony v. Elliott, 5 Colo., 373, 379 (1880).

3. In this case the general scope and purpose of the act of 1879 (L. '79, p. 94, etc.) are stated at length.

Section 2410. . Referee's notice—Contents—How published—Posting copies. Said referee shall prepare and publish a notice containing a copy of the order appointing him, in which notice he shall appoint a time or times, and place or places, suitable and convenient for the claimants in such water district, at which he will attend for the purpose of hearing and taking evidence touching the priority of right of the several ditches, canals and reservoirs in said district, and notifying all persons, associations and corporations interested as owners or consumers of waters (water) to attend by themselves, their agents or attorneys, at the times and places appointed in said notice, and notifying such owners to then and there file a statement of claim in case such statement has not already been filed under the provisions of section one hereof, such as mentioned in section six hereof, and present their proofs touching any priority of right claimed by them for any ditch, canal or reservoir in said district, which notice shall be published in the same manner and times, and in all respects according to the provisions for publication of newspaper notices mentioned in section six of this act, and proof of such publication shall be made in same manner as is provided in section seven of this act ; and he shall also post ten or more printed copies of such notice in ten or more public places in said district, which copies shall be so posted at least twenty days before the time of commencing to take such evidence. L. '81, p. 150, sec. 11; G. S. '83, p. 577, sec. 1773.

1. Sections 1, 6, and 7 referred to are sections 2400, 2405 and 2406 hereof.

Section 2411. Proof of posting notices. Proof of the posting of said copies shall be made by affidavit of said referee or other person certified by him to be a credible witness, which shall show when, where and how the said copies were posted, and shall be filed by him with his report. L. '81, p. 151, sec. 12; G. S. '83, p. 577, sec. 1774.

DIVISION IX.

PROCEEDINGS BEFORE REFEREE.

Section 2412. Who may offer evidence—Former evidence. Said referee shall attend at the times and places mentioned in his said notice for the purpose therein mentioned, and all persons or associations choosing to do so, and being interested as owners of or consumers of water from any ditch, canal or reservoir in said district, and may also attend by themselves, their agents or attorneys, before said referee, at some one or more of said times and places so appointed, and shall have the right to offer any and all evidence they may think advisable for their interests in the matter to be adjudicated, as well in districts in which evidence has been heretofore taken as in other districts. All such evidence as has been heretofore taken, if any, in such district, shall be kept present by said referee, subject to inspection by any party desiring to examine the same for purposes of the investigation. L. '81, p. 151, sec. 13; G. S. '83, pp. 577, 578, sec. 1775.

Section 2413. Powers and duties of referee—Books and records—Evidence. Said referee shall have the power to administer oaths to all witnesses and to issue subpœnas for witnesses and subpœnas *duces tecum*, which subpœnas may be served by any party or constable or sheriff or deputy sheriff, and may require witnesses to appear at any of the places appointed by said referee for taking evidence. He shall permit all witnesses to be examined by the parties calling them, respectively, and to be cross-examined by any party interested, and he shall take all testimony in writing and note all objections offered to any part of the testimony taken, with the cause assigned for the objection, and shall proceed in all other respects as in case of taking depositions. He shall certify all books and papers offered by any one in his own behalf, and preserve them with the testimony offered concerning the said, and in case of books and papers offered in evidence, which shall not be under the control of the party desiring the evidence for which such books may be offered, said referee shall make a true copy of the parts demanded and certify the same, and preserve the same, together with the evidence offered concerning the same and concerning said books and papers, as part of the evidence in the matter. L. '81, pp. 151, 152, sec. 14; G. S. '83, p. 578, sec. 1776.

Section 2414. Refusal to produce books or papers —Effect. No person, association or corporation wilfully refusing to produce any book or paper, if in his or their power to do so, when rightfully demanded for examination and copying, shall be allowed the benefit of any testimony or proofs in his, her or their behalf, in making final adjudication, if the court shall be satisfied, from all the evidence shown concerning such refusal, that the same was wilfull. L. '81, p. 152, sec. 15; G. S. '83, p. 578, sec. 1777.

Section 2415. What facts to be ascertained by proofs. Said referee shall also examine all witnesses to his own satisfaction touching any point involved in the matter in question, and shall ascertain as far as possible the date of the commencement of each ditch, canal or reservoir, with the original size and carrying capacity thereof, the time of the commencement of each enlargement thereof, with the increased carrying capacity thereby occasioned, the length of time spent in such construction or enlargement, the diligence with which the work was prosecuted, the nature of the work as to difficulty of construction, and all such other facts as may tend to show compliance with the law in acquiring the priority of right claimed for such ditch, canal or reservoir; and upon all the facts so obtained shall be determined the relative priorities among the several ditches, canals and reservoirs, the volume or amount of water lawfully appropriated by each, as well as by means of the construction, as by the enlargements thereof, and the time when each such several appropriations took effect. L. '81, p. 152, sec. 16; G. S. '83, p. 578, sec. 1778.

Section 2416. Disturbing proceedings—Penalty. Every person present before said referee at any time when he shall be engaged in hearing testimony, who shall wilfully disturb the proceedings; and every person who shall wilfully refuse or neglect to obey any supbœna issued by said referee, when his lawful fees shall be tendered him for his attendance before the referee, shall be guilty of contempt of the court appointing such referee, and on complaint under oath of the referee or other person, before the said district court, or judge thereof in vacation, may be brought before the court or judge and dealt with accordingly. L. '81, pp. 152, 153, sec. 17; G. S. '83, pp. 578, 579, sec. 1779.

Section 2417. Fees of witnesses—By whom paid. Every witness who shall attend before said referee under subpœna by request of any party, shall be entitled to the same fees and mileage as witnesses before the district court in the county in which he shall so attend, and shall be paid by the party requiring his testimony. L. '81, p. 153, sec. 18; G. S. '83, p. 579, sec. 1780.

Section 2418. Duties of referee—Rights of parties —Adjournment—Notice. The said referee shall take all the testimony offered, and for that purpose shall give reasonable opportunity to all parties to be heard, and may at any place, when the time limited thereat shall expire, adjourn the further taking of testimony then proposed or desired to be offered to the next place in order, according to his said published appointments, and at the last place may continue until all testimony shall be taken, or make further appointment at any former place or places as may seem best and most convenient for all parties, giving reasonable notice thereof. L. '81, p. 153, sec. 19; G. S. '83, p. 579, sec. 1781.

Section 2419. Referee shall examine all testimony—Numbering—Findings—Decree—Report. Said referee, upon closing the testimony, shall proceed to carefully examine the same, together with all testimony and proofs which may have been heretofore taken by any former referee in the same district, if any such shall have been taken, under the provisions of said act, the title of which is recited in section four of this act. He shall make an abstract of all the testimony and proofs in his possession concerning each ditch, canal and reservoir separately, and shall number each ditch and canal in order, and likewise each reservoir, each class consecutively, and also number the several appropriations of water shown by the evidence, all in manner and form as provided in section nine hereof; and shall make a separate finding of all the facts connected with each ditch, canal and reservoir, touching which evidence shall have been offered; and he shall prepare a draft of a decree in accordance with his said findings, in substance the same as the decree mentioned in section four of this act, and conformable also to the provisions of section nine hereof, so far as the same are applicable; which decree, so prepared by him, shall be returned with his report to the court; and he shall file his report with said evidence, abstract and findings, and said decree, with the clerk of the court, and inform the judge of so doing, without delay. L. '81, pp. 153-154, sec. 20; G. S. '83, p. 579, sec. 1782.

1. Sections four and nine referred to herein are respectively sections 2403 and 2408 hereof.

2. Instance of referee's proceedings. Dorr v. Hammond, 7 Colo. 80 (1883).

Section 2420. Filing report—Court proceed to determine—Exceptions—Approval—Entry. Upon the filing of said report the court, or judge thereof in vacation, shall cause an order to be entered setting some day in a regular or special term of court as soon as practicable, when the court shall proceed to hear and determine the report; at which time any party interested may appear by himself or counsel and move exceptions to any matter in the findings or decree made by said referee; and after hearing the same the court shall, if the decree reported be approved, cause the same to be entered of record, or otherwise, such modifications thereof or other decree as shall be found just and conformable to the evidence and the true intent of this act, and to so much of any and all former laws of the state as shall be adjudged consistent therewith. L. '81, p. 154, sec. 21; G. S. '83, pp. 579-580.

1. Referee's judgment on weight of evidence may be reviewed. Dorr v. Hammond, 7 Colo., 80 (1883).

· DIVISION X.

GENERAL PROVISIONS.

Section 2421. Failure to offer evidence—Water commissioner disregard claims until, etc. — Party obtain decree and present certificate. No claim of priority of any person, association or corporation on account of any ditch, canal or reservoir, as to which he, she or they shall have failed or refused to offer evidence under any adjudication herein provided for or heretofore provided for by said act, the title of which is recited in section four hereof, shall be regarded by any water commissioner in distributing water in times of scarcity thereof, until such time as such party shall have by application to the court having jurisdiction, obtained leave and made proof of the priority of right to which such ditch, canal or reservoir shall be justly entitled, which leave shall be granted in all cases upon terms as to notice to other parties interested, and upon payments of all costs, and upon affidavits or petitions sworn to, showing the rights claimed, and the ditches, canals or reservoirs, with the names of the owners thereof against which such priority is claimed, nor until a decree adjudging such priority to such ditch, canal or reservoir has been entered, and certificate, such as mentioned in section four hereof, shall have been issued to claimant and presented to the water commissioner. L. '81, pp. 154, 155, sec. 22; G. S. '83, p. 580, sec. 1784.

1. Section 4 referred to is sec. 2403 hereof.

2. Testimony in former cases when available, sec. 2426.

Section 2422. Rights of parties against referee for neglect, oppression, etc. Every party interested shall have the right to complain to the court of any act of wilfull neglect or oppression on the part of the said referee in exercising his powers under this act, whereby such party shall have been agrieved, either by refusal of said referee to hear or take evidence offered, or by preventing reasonable opportunity to offer such evidence; and the court may order such proceedings in the premises as will give redress of the grievance, at the cost of said referee, if he appear wilfully in fault, otherwise in case of accident or mistake, costs will be awarded as to the court shall seem just. L. '81, p. 155, sec. 23; G. S. '83, p. 580, sec. 1785.

Section 2423. Power of court to make just rules—Law construed liberally. The district court, or judge thereof in vacation, shall have power to make all orders and rules consistent with this act which may be found necessary and expedient from time to time during the progress of the case for carrying out the intent of this act, and of all parts **consistent** therewith of the said act, the title of which **is recited** in section four hereof; as well touching the proceedings in **court as** of the acts and doings of said referee, for the purpose **of** securing **to any** party aggrieved by the acts of said **referee, or any** proceeding of the court, opportunity for redress; **and this** act shall be construed liberally in all courts in favor of securing to all persons interested the just determination and protection of their rights. L. '81, p. **155, sec. 24;** G. S. '83, pp. 580, 581, sec. 1786.

　　1.　Section 4 referred to is sec. 2403 hereof.

　　2.　This section **referred to in Golden Canal Co. v.** Bright, 8 Colo., **155 (1884).**

　　Section 2424.　Party must file claim before offering **evidence.**　No persons, association or corporation representing any ditch, canal or reservoir shall be permitted to give or offer any evidence before said referee until he, she **or** they shall have filed a statement of claim in substance the same **in** all respects as is required to be filed under the provisions of section **one** hereof. L. '81, p. 155, sec. 25; G. S. '83, p. 581, sec. 1787.

　　1.　Section 1 referred to is sec. **2400 hereof.**

　　2.　Filing statement of claim. **Sec. 2400.**

　　Section 2425.　**Re-argument—Review—Limitation** two years.　The **district** court, or judge thereof in vacation, shall have power to order for good cause shown, **and** upon terms just **to all** parties and in such manner as may seem meet, a re-argument **or** review, with or without additional evidence, of **any decree** made under the provisions of this act, **whenever said** court or judge shall find from the cause **shown for that purpose** by any party or parties feeling aggrieved **that the ends** of justice will be **thereby** promoted; **but no** such review or re-argument **shall** be ordered unless applied for by petition or otherwise within **two** years from the **time** of entering the decree complained of. L. '81, p. 156, sec. 26; G. S. '83, p. 581, sec. 1788.

Section 2426. Testimony. Whenever testimony shall or may be taken in any district created by this act, for the purpose of procuring decree as to appropriation of water and priorities thereof, under the statutes of this state, any testimony theretofore taken before any former referee may be introduced, and shall be received as evidence. L. '85, p. 259, sec. 28.

DIVISION XI.

APPEALS.

Section 2227. Who may appeal—Statement—Approval—Contents—Order—Bond—Conditions. Any party or parties representing any ditch, canal or reservoirs, or any party or parties representing two or more ditches canals or reservoirs, which are affected in common with each other by any portion of such decree, by which he or she or they may feel aggrieved, may have an appeal from said district court to the supreme court, and in such case the party or parties joining, desiring an appeal, shall be the appellants, and the parties representing any one or more ditches, canals or reservoirs affecting in common adversely to the interests of appellants shall be the appellees. The party or parties joining in such appeal shall file a statement in writing, verified by affidavit properly entitled in such cause in the district court, which statement shall show that the appellants claim a valuable interest in the ditch, canal or reservoir, or two or more of such, which are affected in common with each other by some portion of said decree, also stating the name or names or otherwise the description of the same, and the name or names or otherwise the description of any one or more other ditches, canals (or) reservoirs, which by said decree derive undue advantage in respect of priority as against that or those represented by appellants; and also setting forth the name or names of the party or parties claiming such other one or more ditches, canals or reservoirs affected in common by said decree adversely to the interest of appellants, and praying that an appeal be allowed against such other parties as appellees. If the court or judge in vacation on examination, find such statement in accordance with the statements of claim filed by the parties named as appellees, mentioned in section one of this act, he shall approve the same and make an order to be prepared and presented by the appellants allowing the appeal and showing the name or names of the appellants and appellees, with the name or names or description of the one or more ditches, canals or reservoirs claimed by the party or parties appellant and appellee, as shown by their several statements of claim filed as aforesaid, before the taking of testimony, and fixing the amount of the appeal bond which bond shall be executed by one or more of appellants, as principal or principals, and by sufficient securities, and approved by the court or judge in vacation, and shall be conditioned for the payment of all costs which may be awarded against the appellants or any of them in the supreme court: L. '81, pp. 156, 157, sec. 27; G. S. '83, pp. 581, 582, sec. 1789.

1. Section 1 of this act referred to last above is sec. 2400 hereof.

2. This section referred to in Golden Canal Co. v. Bright, 8 Colo., 155 (1884).

3. Instance of appeal from report of referee. Dorr v. Hammond, 7 Colo., 80 (1883).

4. Where proofs are mainly taken by a master or referee it is the duty of the supreme court to sift and weigh all the evidence with a view to a just determination, uninfluenced by the proposition that the court below had superior facilities to judge of the credibility of witnesses. Sieber v. Frink, 7 Colo., 152 (1883); Miller v. Taylor, 7 Id., 45 (1881); Jackson v. Allen, 4 Id., 268 (1878); Bates v. Wilson, 24 Pac. Rep., 99 (1890); 14 Colo.

5. But this rule does not apply where the case was heard on testimony taken at a prior trial and also upon parol testimony at the hearing. Bergh v. Rominger, 24 Pac. Rep., 1047 (1890); 14 Colo.

Section 2428. Copy of order served on appellees— Publication—Posting copies—Proof. The order last aforesaid shall be entered of record and the appellant or appellants shall cause a certified copy thereof to be served on each of the appellees, by delivering the same to him or her, if he or she may be found, or otherwise serving the same in manner the same as may be at the time provided for serving summons from the district court by the laws then in force, and shall also cause the said order to be published in the same manner as the notices required to be published by the referee mentioned in section eleven of this act, and proof of the publication in any newspaper shall be the same as in case of said referee's notice, and proof of the posting of the ten printed copies in the district shall be by affidavit of the party posting the same, with the certificate of the clerk of the district court appealed from, that the affiant is a known and credible person. L. '81, p. 157, sec. 28; G. S. '83, p. 582, sec. 1790.

1. Section 11 referred to is sec. 2410 hereof.

Section 2429. Transcript to be filed in six months —Bill of exceptions. The appellant or appellants shall file transcript of record of the district court with the clerk of the supreme court at any time within six months after the appeal shall be allowed as aforesaid. Only so much of the decree appealed from, and so much of the evidence as shall affect the appropriations of water claimed by means of the construction or enlargement or re-enlargement of the several ditches, canals and reservoirs mentioned in the order allowing the appeal, need be copied into the bill of exceptions. L. '81, pp. 157, 158, sec. 29; G. S. '83, p. 582, sec. 1796.

Section 2430. Costs in supreme court. The supreme court on dismissal of such appeal, or on affirming or reversing the parts of the decree appealed from, in whole or in part, shall award costs, as in its discretion shall be found and held to be equitable. L. '81, p. 158, sec. 30; G. S. '83, p. 582, sec. 1792.

Section 2431. Supreme court amend or make new decree, or remand with instructions. The supreme court in all cases in which judgment is rendered and any part of the decree appealed from is reversed, and in which it may be practicable, shall make such decree in the matters involved in the appeal as should have been made by the district court, or direct in what manner the decree of that court shall be amended. L. '81, p. 158, sec. 31; G. S. '83, p. 582, sec. 1793.

Section 2432. Filing proof of service and notice —Sixty days—Supreme court makes rules. The said proof of the service and publication of said order allowing the appeal shall be filed with the clerk of the supreme court within sixty days after the making of said order, and if not so filed the supreme court shall, on motion of the appellee or any of the appellees, at any time after such default in filing said proof and before the said proof shall be filed, dismiss such appeal, and if the transcript of record be not filed within the time limited by sec. 29 of this act such appeal shall, on motion, be dismissed. After the filing of the record and proof of service, aforesaid, the cause on appeal shall be proceeded with as the rules of the supreme court, or such special rules as said court may make in such cases, and their order from time to time thereunder may require. Said court shall have power to make any and all such rules concerning such appeals as may be necessary and expedient in furtherance of this act, as well as to preparation of the case for submission, as to supplying deficiencies of record, if any, and for avoiding unnecessary costs and delay. L. '81, p. 158, sec. 32; G. S. '83, pp. 582, 583, sec. 1794.

1. Section 29 referred to is sec. 2429 hereof.

Section 2433. Court may dismiss referee—Vacancy —New appointment. The district court, or judge thereof in vacation, in case of the death, resignation, absence or other disability of the referee hereby provided for, or for any misconduct in him, or other good cause to such judge appearing, shall appoint such other properly qualified person in his stead as he shall deem proper, who shall proceed without delay to perform all the duties of his office, as herein pointed out, which shall remain unperformed by his predecessor in office. L. '81, p. 159, sec. 33; G. S. '83, sec. 583, sec. 1895.

Section 2434. Suits must be brought in four years—Injunctions in what cases—What districts—Commissioner's duty. Nothing in this act or any decree rendered under the provisions thereof shall prevent any person, association or corporation, from bringing and maintaining any suit or action whatsoever hitherto allowed in any court having jurisdiction, to determine any claim of priority of right to water by appropriation thereof for irrigation or other purposes, at any time within four years after the rendering of a final decree under this act in the water district in which such rights may be claimed, save that no writ of injunction shall issue in any case restraining the use of water for irrigation in any water district wherein such final decree shall have been rendered, which shall effect [affect] the distribution or use of water in any manner adversely to the rights determined and established by and under such decree; but injunctions may issue to restrain the use of water in such district not affected by such decree, and restrain violations of any right thereby established; and the water commissioner of every district where such decree shall have been rendered shall continue to distribute water according to the rights of priority determined by such decree, notwithstanding any suits concerning water rights in such district, until any suits between parties the priorities between them may be otherwise determined, and such water commissioner have official notice by order of the court or judge determining such priorities; which notice shall be in such form and so given as the said judge shall order. L. '81, p. 159, sec. 34; G. S. '83, p. 583, sec. 1796.

Section 2435. **After** four years suit barred. After the lapse of four years from the time of rendering a final decree in any water district, all parties whose interests are thereby affected shall be deemed and held to have acquiesced in the same, except in case of suits before then brought; and thereafter all persons shall be forever barred from setting up any claim of priority of rights to water for irrigation in such water district adverse or contrary to the effect of such decree. L. '81, p. 160, sec: 35; G. S. 83, pp. 583-584, sec. 1797.

Section 2436. Compensation of referee—How paid —His duty to keep account. The referee appointed (as provided) in this act shall be paid the sum of six dollars per day while engaged in discharging his duties as herein provided, and also his reasonable and necessary expenses and mileage at the rate of ten cents for each mile actually and necessarily traveled by him in going and coming in the discharge of his duties as such referee, which said per diem allowance, expenses and mileage shall be paid out of the treasury of the county in which such water district shall lie, if it be contained in one county; and, if such water district shall extend into two or more counties, then, in equal parts thereof, shall be paid out of the treasury of such county into which such district shall extend. He shall keep a just and true account of his services, expenses and mileage, and present the same from time to time to the district court, or judge in vacation, verifying the same by oath, and the judge, if he find the same correct and just, shall certify his approval thereof thereon, and the same shall thereupon be allowed by the board of county commissioners of the county in which said water district shall lie; but, if said (water district) extend into two or more counties, he shall receive from the clerk of the district court separate certificates, under seal of the court, showing the amount due him from each county, upon which certificate the board of county commissioners of the respective counties shall allow the same on presentation thereof. L. '81, p. 160, sec. 36; G. S. '83, p. 584; sec. 1798.

Section 2437. Repeal. All laws and parts of laws heretofore in existence inconsistent with the provisions of this act, shall be and the same are hereby repealed. L. '81, p. 160, sec. 37; G. S. '83, p. 584, sec. 1799.

Section 2438. Sheriff not serve writ outside his county. Nothing herein contained shall be construed to authorize any sheriff to serve any writ outside of the limits of his own county, or give effect to any record by way of notice or otherwise, in any county other than

that in which it belongs.   L. '79, p. 106, sec. 31; G. S.
'83, p. 584, sec. 1800.   •

Section 2439.   Fees of district clerk—How audited
—Paid.   The fees of the clerk of the district court for
a service rendered under this act shall be paid by the
counties interested in the same manner as the fees of
the water commissioners, upon said clerk rendering his
account, certified by the district judge to the board or
boards of county commissioners of the county or
counties embracing the water district is case of which
the service shall have been rendered.   L. '79, p. 108, sec.
43; G. S. '83, p. 584, sec. 1801.   •

DIVISION XII.

WATER DIVISIONS.

Section 2440.   Water division constituted.   That
for the better regulating of the distribution of water for
irrigation among the several ditches, canals and reservoirs into which such water may be lawfully taken in
times of scarcity thereof, the water districts now, or to
be hereafter, established by law, shall be constituted
into water divisions as follows: L. '81, p. 119, sec. 1;
G S. '83, p. 585, sec. 1802.

Section 2441.   Water division No. 1.   That all
water districts now or hereafter to be formed, consisting
of lands in the state of Colorado irrigated by water taken
from the South Platte river, the North Platte river, the
Big Laramie river, the north and middle forks of the
Republican river, Sandy and Frenchman's creeks, and
the streams draining into the said rivers and creeks,
shall constitute water division No. 1.   L. '81, p. 119,
sec. 2; G. S. '83, p. 585, sec. 1803; amended L. '89, p.
211, sec. 8.   •

Section 2442.   Water division No. 2.   Arkansas division.   "That all water districts now or hereafter to be
formed, consisting of lands irrigated by water taken
from the Arkansas river, the south fork of the Republi-

can river, the **Smoky Hill river and the Dry Cimarron river, and the streams draining into the said rivers, shall** constitute water division **No. 2 and** be named **the Arkansas division."** L. **'81, p. 119, sec. 2**; G. S. '83, p. 585, sec. **1804**; amended L. **'89, p. 472, sec. 4.**

Section 24**4**3. Water division **No.** 3. Rio **Grande** division. All **water** districts now or hereafter **to be** formed, consisting of lands watered from the Rio **Grande** river and its tributaries, shall constitute water **division** No. 3 and be named the Rio Grande division. L. **'81,** p. 119, sec. 4: G. S. '83, p. 585, sec. 105.

Section 2444. Water division No. 4. San Juan division. That all water districts now, or hereafter to be formed, consisting of lands in the state of Colorado watered by the San Juan river and its tributaries, shall constitute water division number four (4) and be named the San Juan division. L. '85, p. 256, sec. 2.

Section 2445. **Water division No.** 5, Grand **river** division. That all water districts **now** or hereafter **to** be formed, consisting of lands in **the** state of Colorado watered by the Grand river **and its** tributaries, shall constitute water division No. **5,** and be named **the** Grand river division. L. '87, p. 313, sec. 1.

Section 2446. Water division No. 6, Green **river** division. **That all** water districts now or hereafter to be **formed, consisting of** lands in the state of Colorado irri-**gated by water taken** taken from the Green river and **its tributaries, shall** constitute water division No. 6, and **be named the Green river** division. L. '89, p. 210, **sec. 1.**

Section 2447. **Superintendents of** irrigation. That the governor shall **appoint a** superintendent of irrigation for each of the water **divisions now** existing within in **the** state, or which may **hereafter** be created, such superintendents of irrigation to **hold office** for a period of two years from the date of their respective appointments, **or until** their successors shall be appointed and qualified. **The** governor may at any time in his discretion remove said superintendents of irrigation, or any of them, and appoint others in their stead, for the remainder of said term of two years; *Provided*, That the governor shall not appoint a superintendent of irrigation in any district until the board of county commissioners of some one or more of the counties whose territory, or any part of whose territory is included in such water district, shall have, at a meeting regularly called and held, adopted a resolution requesting such appointment to be made, and have had the same certified to the governor. L. '87, p. 295, sec. 1.

1. He is superior to water commissioners, sec. 2386.

Section 2448. Duties of superintendent. Said superintendent of irrigation shall have general control over the water commissioners of the several districts within his division. He shall, under the general supervision of the state engineer, execute the laws of the state relative to the distribution of water, in accordance with the right of priority of appropriation, as established by judicial decree, and perform such other functions as may be assigned to him by the state engineer. L. '87, p. 296, sec. 2.

Section 2449. Superintendents may make regulations. Said superintendent of irrigation shall, in the distribution of water, be governed by the regulations of this act, and acts that are now in force; but for the better discharge of his duties, he shall have the authority to make such other regulations to secure the equal and fair distribution of water, in accordance with the rights of priority of appropriation, as may in his judgment be needed in his division; *Provided*, Such regulations shall not be in violation of any part of this act, or other laws of the state, but shall be merely supplementary to and necessary to enforce the provisions of the general laws and amendments thereto. L. '87, p. 296, sec. 3.

Section 2450. Appeal to state engineer. Any person, ditch company or ditch owner, who may deem himself injured or discriminated against by any such order or regulation of such superintendent of irrigation shall have the right to appeal from the same to the state engineer by filing with the state engineer a copy of the order or regulation complained of, and a statement of the manner in which the same injuriously affects the petitioner's interest. The state engineer shall after due notice, hear whatever testimony may be brought forward by the petitioner, either orally or by way of affidavits, and through the superintendent of irrigation, shall have power to suspend, amend or confirm the order complained of. L. '87, p. 296, sec. 4.

Section 2451. When duties shall commence—When close—Report—Compensation. Said superintendent of irrigation shall commence the discharge of his duties in his division as soon as the first water commissioner in any district within his division shall be called out, and shall continue to discharge his duties until the last water commissioner in any district of his division ceases to be needed. Each water commissioner shall report immediately to the superintendent of irrigation of his division when he is called out and when he ceases to be needed, and shall during the continuance of his duties, be under the control of the superintendent of irrigation of his division. The superintendent of irrigation shall receive, as compensation, five dollars per day for every day during which he is employed in the discharge of his duty. L. '87, p. 296, sec. 5.

Section 2452. Bond of superintendent. Within thirty days after the appointment of said superintendent of irrigation it shall be his duty to give bond to the amount of five thousand dollars for the faithful discharge of his duty, said bond to be approved by the board of county commissioners of the county wherein said superintendent of irrigation may reside, and to be filed in the office of the county clerk and recorder of such county. L. '87, p. 297, sec. 6.

Section 2453. Clerk of district court furnish copies of decrees—Register of priorities—tabulated statement. Within thirty days after his appointment said superintendent of irrigation shall send to the clerk of the district court, within his division of such counties as have had rendered by the district court of such county judicial decrees fixing the priorities of appropriation of water for irrigation purposes for any water district, a notification of his appointment to such office, and shall request of the said clerk a certified copy of every decree of the district court establishing priorities of appropriation of water used for irrigation purposes within that district.   Thereupon, it shall be the duty of such clerk, within ten days after receipt of such request from said superintendent of irrigation, to prepare a certified copy of all decrees of such district court establishing priorities of water rights made within that district under the provisions of the general statutes of the state of Colorado, and transmit the same to the superintendent of irrigation requesting it.   Said superintendent of irrigation shall then cause to be prepared a book to be entitled "The Register of Priorities of Appropriation of Water Rights for Water Division No.____, State of Colorado," within which he shall enter and preserve such certified copies of decrees.   Said superintendent of irrigation shall, from such certified copies of decrees, make out a list of all the ditches, canals and reservoirs entitled to appropriations of water within his division, arranging and numbering the same in consecutive order, according to the dates of their respective appropriations within his division, and without regard to the number of such ditches, canals or reservoirs may bear within their respective water districts.   Said superintendent of irrigation shall make from his register a tabulated statement of all the ditches, canals and reservoirs in his division whose priorities have been decreed, which statement shall contain the following information concerning each ditch, canal and reservoir arranged in separate columns: The name of the ditch, canal or reservoir; its number in his division; the district in which it is situated; the number of it in its proper district, and the number of cubic feet of water per second to which it is entitled, and such other and further information as he may deem useful to the proper discharge of his duty. In case any decrees of court establishing priorities of appropriation of water for irrigation purposes are made after the transmittal of the copy of previous decrees to the superintendent of irrigation, it shall be the duty of the clerk of the court wherein such decree is rendered to transmit to the superintendent of irrigation of the division within which said county is situated, within ten days after it is rendered, a copy of such decree, and the superintendent of irrigation shall enter the same in his register, such register to be filed and kept in the office of the state engineer. L. '89, p. 297, sec. 7.

Section 2454. Superintendent may call out commissioner at any time. Said superintendent of irrigation shall have the right to call out any water commissioner of any water district within his division, at any time he may deem it necessary, and he shall have the power to perform the regular duties of water commissioner in all the districts within his division. L. '87, p. 298, sec. 8.

Section 2455. Reports of commissioners and superintendents—Contents—Enforcement of priority. All water commissioners shall make reports to the superintendent of irrigation of their division as often as it may be deemed necessary by said superintendent. Said report shall contain the following information: The amount of water necessary to supply all the ditches, canals and reservoirs of that district; the amount of water actually coming into the district to supply such ditches, canals and reservoirs; whether such supply is on the increase or decrease; what ditches, canals or reservoirs are at that time without their proper supply; the probability as to what the supply will be during the period before the next report will be required, and such other and further information as the superintendent of irrigation of that division may suggest. Said superintendent of irrigation shall carefully file and preserve such reports, and shall, from them, ascertain what ditches, canals and reservoirs are, and what are not, receiving their proper supply of water, and if it shall appear that in any district in that division any ditch, canal or reservoir is receiving water whose priority post dates that of the ditch, canal or reservoir in another district, as ascertained from his register, he shall at once order such post-dated ditch, canal or reservoir shut down, and the water given to the elder ditch, canal or reservoir. His orders being directed at all times to the enforcement of priority of appropriation, according to his tabulated statement of priorities, to the whole division and without regard to the district within which the ditches, canals and reservoirs may be located. The reports of water commissioners by the superintendents of irrigation shall be filed and kept in the office of the state engineer. L. '87, p. 298, sec. 9.

Section 2456. Owners of ditches report failure to receive supply of water—Apportionment—Report of commissioner. In case any ditch, canal or reservoir in any district within such superintendent of irrigation's division shall fail to receive its regular supply of water, the owner or controller of such ditch, canal or reservoir may report such fact to the water commissioner of that district, who shall immediately apportion the water in his district, and send forthwith by telegram, if necessary, a report of such fact to the superintendent of irrigation of his division, and thereupon it shall be the duty of said superintendent to compare such report with his register, and if any ditch, canal or reservoir of any other district of his division is receiving water to which any ditch, canal or reservoir of any other district is entitled, he shall at once order the shutting down of the post-dated ditches, canals or reservoirs, and the water given to the ditches, canals or reservoirs having the priority of appropriation; *Provided, however,* That nothing in this act shall be construed as interfering with the priority of water for domestic use. L. '87, p. 299, sec. 10.

Section 2457. Expenses and salary of superintendents—Clerk's fees. The expenses and salary of the superintendents of irrigation shall be paid *pro rata* by the counties interested, in the same manner as the fees of water commissioners are paid, and the fees of the clerks of the district courts, for services rendered under the provisions of this act, shall also be paid by the counties interested, upon the said clerk rendering his account, certified by the district judge, to the boards of county commissioners of the counties embraced in the water divisions in case of which the services have been rendered L. '87, p. 299, sec. 11.

DIVISION XIII.

STATE ENGINEER.

Section 2458. **Governor** to appoint a state engineer—Office—Salary—Oath—Bond. The governor **shall** appoint a state engineer, who shall hold his **office for** the term of two years, or **until his successor shall be** appointed and qualified. **The governor may at any time,** for cause shown, remove **said state** engineer. The said **state** engineer shall have **his office at** the state capitol, **in suitable** rooms, to be provided **for** him by the secretary **of state,** who shall furnish him with suitable furniture, postage and such proper and necessary stationery, books and instruments as are required to best enable him to discharge the duties of his office. He shall be paid a salary **of** three thousand dollars per annum, payable monthly **by** the state treasurer, on warrants drawn **by** the state auditor. The said state engineer shall, **before** entering **on** the discharge of his duties, take and **subscribe** to an oath, before the judge of a state **court of** record, to faithfully perform the duties of his office, **and file** said oath with the secretary of state, together with his official **bond, in the** penal sum of ten thousand dollars; **said bond to be** signed by sureties approved by the **secretary of state and** conditioned upon the faithful **discharge of the duties of** his office, and for delivering to his successor **or other officer** authorized by the governor **to receive the same all** moneys, books, instruments and other property **belonging to** the state then in his possession **or** under his **control, or** with which **he** may be legally chargeable **as such state** engineer. L. '89, p. 371, sec. 1.

1. **He is** superior to water commissioner, sec. 2386.

2. **The** governor may appoint a state engineer **without the advice** and **consent** of the senate. *In re.*

Question by governor, 12 Colo., 400 (1888). See this case in note 11 to sec. 6, art. IV., Const.

Section 2459. Engineer **control** waters, make measurements, collect data. The **state** engineer shall have general supervising control over the public **waters** of the state. He shall make or **cause** to be made careful measurements of the flow of **the** public **streams** of the state from which water is diverted for **any purpose** and compute the discharge of **the same.** He shall also collect all necessary data and **information** regarding the location, size, cost and capacity of dams and reservoirs hereafter to be constructed, and like data regarding the feasibility **and** economical construction of reservoirs on eligible sites, of which he may obtain information, and the **useful** purposes to which the water from the same **may be put.** He shall **also** collect all data and information regarding the snow-fall in **the** mountains each season, for the purpose of predicting **the** probable flow of water in the streams of the state, and publish the same. L. '89, p. 372, sec. 2.

Section 2460. Approve **designs and** plans **for dams** and embankments. The state engineer shall **approve** the designs and plans for the construction and **repair** of all dams or reservoirs, embankments which **are** built within the state, which equal or exceed ten feet **in** vertical height. L. '89, p. **372**, sec. 3.

Section 2461. **State** engineer **to have** general charge of work—**Require** reports. The state engineer shall have general charge over the work of the division water superintendents, and district **water** commissioners, and shall furnish them with all **data** and information necessary for the proper and intelligent discharge of the **duties** of their offices, and **shall** require them **to** report to him at suitable times **their** official actions, **and** require of them annual statements **on** blanks to be furnished by him, of the amount of water diverted from **the** public streams in their respective divisions **and districts**, and such other statistics as, in the judgment **of (the) state engineer** will be of **benefit** to the state. L. '89, p. **373**, sec. 4.

Section 2462. Appoint a deputy to measure. The state engineer shall, on request of any party interested, and on payment of his per diem, charge and reasonable expenses, appoint a deputy to measure, compute and ascertain all necessary data of any canal, dam, reservoir or other construction, as required or as may be desired to establish court decrees, or for filing statements in compliance with law, in the county clerk's records. L. '89, p. 373, sec. 5.

Section 2463. Perform all duties imposed upon him—Expenses. The state engineer shall, without any extra pay or compensation beyond the salary provided in section 1 of this act, perform all duties imposed upon him by law, and shall when called upon by the governor, give his counsel and services, without extra pay or compensation, to any state department or institution; *Provided, however*, That he shall be allowed all actual traveling and other necessary expenses, and the actual cost of preparing necessary maps and drawings, which actual expenses shall be paid by the department or institution requiring his services. L. '89, p. 373, sec. 6.

1. Section 1 referred to herein is sec. 2458 hereof.

Section 2464. May appoint deputies and revoke the same—Bond—Oath. The state engineer may appoint one or more deputies, as he may think proper, for whose official actions he shall be responsible, and may revoke such appointments at his pleasure, and he may also deputize any person to do a particular service; and the said state engineer and his sureties shall be responsible on his official bond for the default or misconduct of his deputies. Such appointment and revocation shall be in writing, under the signature and official seal of the state engineer, and shall be filed in the office of the state secretary of state. All persons appointed shall take and subscribe to an oath before the judge of a court of record to truly perform the duties of the office to which he is appointed, and such oath shall be filed with his appointment in the office of the secretary of state. In addition to the deputies provided for in this section, the state engineer may employ such assistance in performing the work of his office as he may deem necessary. L. '89, p. 373, sec. 7.

Section 2465. Pay of deputies and assistants. **The** pay of the deputies and assistants of the state engineer shall not exceed the sum of six dollars per day for each day employed, together with the actual expenses, and the whole amount which may be so expended is hereby limited to the sum of forty-five hundred dollars each year. L. '89, p. 374, sec. 8.

Section 2466. Require owners **of ditches to con-struct** and maintain **a** measuring weir. **For the** more **accurate and convenient** measurement of any water ap-**priated pursuant to any** judgment or decree rendered by **any court establishing** the claims of priority of any ditch, **canal or reservoir, the** owners thereof may be required **by the state** engineer to construct and maintain, under **the** supervision of the **state** engineer, a measuring weir or other device for measuring the flow of the water at **the** head of such ditch, canal or reservoir, or as near thereto **as** practicable. The state engineer shall compute and arrange in tabular form the amount of water that **will** pass such weir or measuring device at the different stages thereof, **and** he shall furnish a copy of a statement thereof **to any** water superintendents or commissioners having **control** of such ditch, canal or reservoir. L. '89, p. 374, sec. 9.

Section 2467. **Unit** of measurement. The state engineer shall use in **all his** calculations, measurements, records and reports, the cubic foot per second as the unit of measurement of flowing water, and the cubic foot as the unit **of** measurement of volume. L. '89, p. 374, sec. 10.

1.   Measure by statute inch.   Sec. 4643.

Section 2468. Engineer **prepare** true report. The state engineer shall prepare and render to the governor a full and true report of his **work**, regarding all matters and duties devolving upon him by virtue of his office, which report **shall** be delivered at the time when the reports of **other state** officers are required by law to be made, in order that it may be laid before the general assembly at each regular session thereof. L. '89, p. 375, sec. 11

Section 2469. Repeal. Sections six, seven, eight, nine, ten, eleven and twelve of an act entitled "An act to provide for the appointment of a state engineer, and to define his duties and regulate his pay, and for the appointment of his assistants and the establishment of water divisions," approved March 5, 1881, the same being general section eighteen hundred and seven, eighteen hundred and eight, eighteen hundred and nine, eighteen hundred and ten, eighteen hundred and eleven, eighteen hundred and twelve, eighteen hundred and thirteen of the general statutes of the state of Colorado, are hereby repealed. L. '89, p. 375, sec. 12.

DIVISION XVI.

STATE CONTROL.

Section 2470. Coal creek reservoir—Appropriation—Proviso. There is hereby appropriated out of any money in the state treasury belonging to the internal improvement permanent fund, and any money which may hereafter be credited to said fund and not otherwise appropriated, the sum of twenty thousand (20,000) dollars, or as much thereof as may be necessary, as is hereinafter provided, for the construction of a reservoir at Coal creek, upon or adjacent to sections twenty, twenty-eight or thirty-four, township four, south range sixty-five west, in the county of Arapahoe, to store the water of floods for the purpose of irrigation and other beneficial uses; *Provided*, That no part of said appropriation shall be used for the purchase of land, and that the said reservoir shall not be constructed except upon lands the title to which shall be first re-vested in the state; and, *Provided further*, That all citizens of the state shall have free and equal rights to the use and benefits of said reservoir when constructed, subject only to such reasonable rules and restrictions as may be provided by law for the protection of the property. L. '89, p. 215, sec. 1.

Section 2471.   State engineer measure water—Determine capacity—Prepare plans.   As soon as practicable after the passage and approval of this act the state engineer shall make the necessary arrangements for measuring the flow of water in said Coal creek, with a view of constructing a reservoir of sufficient capacity to hold the waters that may result from storms in that portion of the state drained by said Coal creek and above said reservoir.   Said state engineer shall thereafter calculate and determine the required capacity of a reservoir to store the waters flowing in said creek, and prepare plans and specifications thereof.   L. '89, p. 216, sec. 2.

Section 2472.   Construction board—Private donations.   The governor, state engineer and attorney general shall be, and hereby are constituted a board for the purpose of constructing said reservoir and taking charge of the same until otherwise provided by law; *Provided*, That if, after proper examination and survey, the board shall determine that it is not practicable and feasible to construct said reservoir at the place herein designated, or that the same cannot be properly constructed with the sum appropriated by this act, together with such private donations and subscriptions as may be tendered to the board, then no portion of said appropriation shall be expended, except so much as may have been necessary to defray the expenses of such examinations and survey.   L. 89, p. 216, sec. 3.

Section 2473.   Plans and specifications—Advertise for bids.   Upon the preparation of the plans and specifications by the state engineer, it shall be the duty of the said board to advertise for bids in accordance therewith, and thereupon they shall let the contract to the lowest responsible bidder.   L. '89, p. 216, sec. 4.   •

Section 2474.   Warrants for expenses.   The auditor of state is hereby authorized to draw warrants for the payment of the expenses of building said reservoir, upon vouchers certified to by the aforesaid board, not exceeding the said sum of twenty thousand (20,000) dollars. L. '89, p. 217, sec. 5.

Section 2475. Water-works property of state—Engineer provide for delivery of water. That the said reservoir and water-works, and the waters when so collected and stored, shall be the property of the state, and the water so supplied shall be turned into Coal creek or canal for the purpose of supplying water for appropriations heretofore made, or hereafter to be made, in the order of such appropriation by the several canals and reservoirs taken from said stream. The state engineer, or in his stead such person or persons as may be duly appointed for that purpose according to law, shall determine, regulate and provide for the delivery of such water to such ditches, canals and reservoirs, according to their several appropriations, decrees of court, capacities and necessities. L. '89, p. 217, sec. 6.

Section 2476. Shall not impair acquired rights. Nothing in this act shall be construed so as to impair any rights acquired, or that may be acquired under or by virtue of the irrigation laws of the state of Colorado. L. 89, p. 217, sec. 7. ·

Section 2477. Damaging reservoir a misdemeanor —Penalty. Any person interfering with or damaging said reservoir or any of its approaches or appurtenances, shall be deemed guilty of a misdemeanor, and upon conviction thereof, shall be fined not exceeding one thousand (1000) dollars, or by imprisonment in the county jail not exceeding one year. L. '89, p. 217, sec. 8.

Section 2478. Appropriation for preliminary survey of the Grand, Laramie and North Platte. That there is hereby appropriated out of any funds in the state treasury belonging to the internal improvement fund not otherwise appropriated the sum of three thousand dollars, or so much thereof as is necessary to defray the necessary expenses of a preliminary survey and investigation of the sources of the Grand, Laramie and North Platte river systems, with reference to turning the unappropriated waters thereof eastward and causing

them to flow into and through the tributaries of the
South ⌐latte and Arkansas river systems for the purpose
of irrigation and other beneficial uses. L. '89, p. 208,
sec. 1.

Section 2479.   Competent engineers to be employed
for surveys and estimates—Report.   That the governor,
attorney general and state engineer are hereby authorized
to employ a competent engineer or engineers, and to
cause a survey to be made of the sources of the Grand,
Laramie and North Platte river systems, and the tributaries thereto, at or near the continental divide, and to
determine whether the unappropriated waters thereof
can be made to flow eastward into and through the
South Platte and Arkansas river systems, as aforesaid,
and to determine the practicability and feasibility of
such diversion, and the means necessary to be used to
secure the same, together with an estimate of the probable cost in detail and severally, of diverting the unappropriated waters from such stream or streams as they
may find can be so diverted; and the person or persons
to whom such work is committed, as aforesaid, shall report thereon to the governor, attorney general and state
engineer in form and manner as above provided. L. '89,
p. 209, sec. 2.

Section 2480.   If diversion feasible, plans and
specifications to be prepared and contracts let.   If it be
found from the report of such survey that the diversion
of waters as aforesaid is feasible and practicable, and
will be beneficial to the state, the governor, attorney
general and state engineer, acting as a commission in
that behalf, shall proceed without unecessary delay to
obtain plans and specifications in relation to the several
diversions proposed and determined to be practicable,
and thereupon shall let contracts to construct ditches,
canals, dams and water works for such purpose from any
and all of such streams and tributaries belonging to the
aforesaid Grand, Laramie and North Platte systems,
which in their judgment they may deem expedient, and
all such contracts and works to be done and performed
under the supervision of said commission, and according to plans and specifications by them adopted in relation thereto. L. '89, p. 209, sec. 3.

Section 2481. Waters and works property of state
—How water applied. That the said ditches, canals
and water works, and the waters when so diverted,
shall be the property of the state, and the waters so
supplied shall be turned into the said South Platte and
Arkansas rivers and their tributaries for the purpose of
supplying deficiencies of water for appropriation heretofore made, or hereafter to be made, in the order of
such appropriation by the several canals and reservoirs
taken from said streams. The state engineer, or in his
stead such person or persons as may be duly appointed
for that purpose according to law, shall determine, regulate and provide for the delivery of such waters to
such ditches, canals and reservoirs, according to their
several appropriations, decrees of court, capacities and
necessities. L. '89, p. 210, sec. 4.

Section 2482. Appropriation for expenses. There
is hereby appropriated out of any funds in the treasury
not otherwise appropriated and belonging to the general internal improvement fund, the sum of ten thous-
and dollars to pay for the construction of such canals,
ditches and water works as may be so determined upon,
and to defray the necessary expenses of such enterprise;
said sum, or so much thereof as is necessary to be drawn
as the work progresses, and as necessity may require,
upon the order of said commission drawn upon the state
auditor, who shall issue his warrant therefor upon the
state treasury; all bills shall be audited and and
approved by said commission only. L. '89, p. 210, sec. 5.

Section 2483. Penitentiary commissioners may use
convict labor to construct ditches. That for the purpose of reclaiming by irrigation, state and other lands,
and for the purpose of furnishing work for the convicts
confined in the state penitentiary, the board of commissioners of the state penitentiary is hereby authorized to
locate, acquire and construct in the name of, and for the
use of the state of Colorado, ditches, canals, reservoirs
and feeders, for irrigating and domestic purposes, and
for that purpose may use convict labor of persons confined, or that may be confined, as convicts in the state
penitentiary at Canon City L. '89, p. 285, sec. 1.

Section 2484. State engineer survey ditches from Arkansas river. The state engineer, under the direction of the board, shall survey, lay out and locate a ditch or canal upon the most feasible route on either side of the Arkansas river, which said ditch or canal shall be of sufficient capacity to cover at least thirty thousand acres of good arable land between Canon City and Pueblo; *Provided*, That work shall only be commenced and performed upon one main ditch, canal, reservoir or feeder at a time; that a second shall not be commenced until the completion of the first. L. '89, p. 285, sec. 2.

Section 2485. Board given all rights and powers. The said board is hereby given all the rights and powers that an individual or corporation now has, or may hereafter have, under the laws of the state, or of the United States, to acquire the right of way over, upon, and to any lands necessary for it to use or occupy in the construction and maintenance of said ditches, canals, reservoirs or feeders. L. '89, p. 286, sec. 3.

Section 2486. Title shall vest in state. That the title to all ditches, canals, reservoirs or feeders so constructed under this act shall vest and remain in the state of Colorado, and the proceeds thereof shall be paid into the state treasury. L. '89, p. 286, sec. 4.

Section 2487. Contract for and lease water rights. That when any part of any ditch, canal, reservoir or feeder shall be constructed under this act, said board of penitentiary commissioners may contract for and may lease water rights upon such terms and under such rules and regulations as may be adopted by said board and approved by the governor of the state to such individuals or corporations as may desire to lease the same. L. '89, p. 286, sec. 5.

Section 2488. Certificates issued for subscriptions draw interest—How payable. That for the purpose of aiding in the construction of said ditches, canals, reservoirs and feeders, the said board is hereby authorized to receive subscriptions and advancements of money from persons owning land along the line of said proposed ditches, canals, reservoirs and feeders, or persons desiring the construction of the same, and to issue receipts or certificates to such person or persons so advancing money for the amount thereof, which receipt or certificate shall draw interest at the rate of seven per cent. per annum, and both principal and interest shall be payable in water to be taken from said ditches, canals, reservoirs or feeders, under such rules and regulations as may be adopted by said board and the state engineer and approved by the governor of the state. L. '89, p. 286, sec. 6.

Section 2489. Appropriation of materials. There is hereby appropriated out of any money in the state treasury not otherwise appropriated, for the purpose of locating and paying for powder, fuse, tools, teams and material used in the construction of said ditches, canals, reservoirs and feeders, as provided for in this act, the sum of ten thousand dollars. L. '89, p. 286, sec. 7.

Section 2490. Repeal. All acts or parts of acts inconsistent with the provisions of this act are hereby repealed. L. '89, p. 287, sec. 8.

Section 2491. Appropriation for South Boulder canal—Purpose. There is hereby appropriated out of the fund for internal improvement the sum of twenty-five thousand (25,000) dollars, or so much thereof as may be necessary, for the purpose of making a survey and for the construction of a canal along the western slope of the range for a distance of twenty miles, more or less, and to cut across the range and connect with the South Boulder creek, for the purpose of increasing the supply of water in said South Boulder creek for agricultural purposes. L. '89, p. 46, sec. 1.

Section 2492. Who shall constitute the board of construction. The board of county commissioners of Boulder county and the state engineer shall constitute a board for the purpose of making said survey and locating and constructing said canal. L. '89, p. 46, sec. 2.

Section 2493. Shall be in Water District No. 6. That said survey when made shall be deemed to be in Water District No. 6. L. '89, p. 46, sec. 3.

Section 2494. State treasurer shall sell state warrants to provide funds—Proviso. The state treasurer is hereby authorized and directed to sell state warrants belonging to the internal improvement fund or internal improvement income fund, for the purpose of providing funds for carrying on the work herein provided for, whenever the amount of cash in the treasury belonging to said fund is exhausted. L. '89, p. 46, sec. 4.

Section 2495. Provided such improvement be practicable. *Provided*, That no portion of the funds appropriated be for lands upon which to place such improvements, and no such improvements shall be made if it be found impracticable by the state engineer, after a careful survey has been made, except that all expenses of said surveys shall be paid for out of the internal improvement fund. L. '89, p. 47, sec. 5.

Section 2496. Commission for purification of waters of Clear creek. That there is hereby created a commission, consisting of the state engineer, the president of the faculty of the state school of mines, and the president of the faculty of the state agricultural college, for the purpose of making experiments, and practical tests of the waters of Clear creek in the counties of Gilpin, Clear Creek and Jefferson, with a view to the purification of the waters of said stream. L. '89, p. 311, sec. 1.

1. For experimenting in purifying the waters of the streams, lakes and reservoirs of the state $600.00 was appropriated in 1885. L. '85, p. 260.

Section 2497. Commission use due diligence to devise plan, make tests—State engineer superintendent. That said commission is directed to use due diligence in endeavoring to devise some practical method of arresting the sediment from said stream below the stamp mills located on the same, and looking to the purification of said waters. Said method may be by any new process, plan, in the line of civil engineering or other scheme which, after investigation, the commission may deem feasible, and when a plan to accomplish said purpose is agreed upon by said commission, they are authorized to employ all necessary assistance to have carried out a practical test of said plan so far as the money for said use appropriated in this act may permit. The state engineer shall superintend such work according to the commissioner's direction. L. '89, p. 312, sec. 2.

Section 2498. Compensation of employees. That the employees who may be engaged in the work, under the direction of the state engineer, in carrying out the tests and experiments suggested by the commission shall be entitled to such reasonable compensation, for their services, as the commission may deem proper; *Provided*, That in no case shall it be lawful for the commission or engineer, in carrying out the plan to incur more expense than the money herewith appropriated will liquidate. L. '89, p. 312, sec. 3.

Section 2499. Report fully its proceedings, results, etc. That it shall be the duty of said commission to report fully of its proceedings, expenditures and results to the next session of the general assembly of the state, said report to be made through the state engineer's office. L. '89, p. 312, sec. 4.

Section 2500. Appropriation. That there is hereby appropriated out of any moneys in the treasury not otherwise appropriated the sum of five thousand (5,000) dollars, or so much thereof as may be required to carry out the provisions of this act. L. '89, p. 312, sec. 5.

Section 2501. Auditor of state authorized to draw warrant. That the auditor of state is hereby authorized to draw his warrants upon the state treasurer in payment of said moneys for said expenses incurred, upon vouchers duly approved by said president of the faculty of the state School of Mines and president of the faculty of the state agricultural college, when countersigned by the state engineer. L. '89, p. 312, sec. 6.

Section 2502. Governor appoint commission to draft code of law concerning waters of state. That within ten days after the passage of this act the governor, by and with the advice and consent of the senate shall appoint three persons as a commission, whose duty it shall be to draft and report, for submission to the next general assembly, a complete revision and code of law concerning the waters of the state, as derived from natural streams, springs, artesian wells, drainage, percolation and other sources; prescribing the methods, facilities and appliances for the control, regulation, use and disposition of said waters, and providing for the official management thereof, in form and manner as hereinafter set forth.   L. '89, p. 466, sec. 1.

Section 2503.   Duty of commissioners.   It shall be the duty of said commissioners, upon their acceptance of said appointment, to jointly enter upon the work of drafting, framing, digesting and codifying a complete system of law in accordance with the provisions of the constitution, and subject to rights vested thereunder, embracing the whole subject of the waters of the state; whether such waters are derived from natural streams, springs, surface or underground channels, artesian wells, rainfall, melting snow, flood waters, percolating and seepage waters, water collected by drainage, and from any and every source of accumulation and supply; to provide for the appropriation, regulation, distribution, use and economy of the same for agricultural, domestic, mechanical and mining purposes, by canals, reservoirs, drains, conduits, pipes or otherwise; to provide for the redemption of swamp and seepage lands by drainage, and for the utilization of the water collected for drainage works, and, generally, to formulate a complete system of laws in relation to waters derived and collected from any and every source and used, employed and disposed of in the various and beneficial uses and disposition to which water is applied, under conditions existing in the state, and to provide for the officers, officers' powers and facilities necessary to carry out and enforce the provisions of such system of law.   L. '89, p. 467, sec. 2.

Section 2504.   Limitation of time—Expenses, how paid.   The said commission shall carry on its said work at the time and places and according to the rules and regulations agreed upon among the members thereof, and within the time intervening between its appointment and the first Monday in December next preceding the sitting of the next general assembly, at which time it shall be prepared to report a draft for a code or system of law as provided in section two of this act; and, for the purpose of facilitating its work, such commission shall have power to employ such clerical assistance as, in its judgment, it shall find necessary for carrying on and completion of its duties, and to purchase such books, supplies and materials as shall be necessary to or connected with the work of said commission, at the total expense for said clerical assistance, books and materials and contingent expenses not to exceed four thousand (4,000) dollars, which shall be paid out of any moneys appropriated for that purpose, on certificates signed by said commissioners, showing the services rendered and the amount thereof, and on presentation of such certificate to the state auditor by the person or persons entitled thereto, he shall issue his warrant or warrants on the state treasurer for the amount thereof, to be paid out of any appropriation as aforesaid. L. '89, p. 467, sec. 3.

1. Section 2 referred to is sec. 2503 hereof.

Section 2505. Report of commission. The said commission shall prepare its report to be submitted in writing to the eighth general assembly, which report shall be completed on or before the first Monday in December next preceding the sitting of the said eighth general assembly, and deliver the same to the secretary of state on or before said date; which report shall include the form of a bill to enact the recommendation of the commission into a law, with a proper title for such proposed enactment. If the commission shall fail to agree upon a complete report, all such matters as are not agreed to shall be submitted, as to those matters only, in a minority report or reports, such minority report or reports to be also delivered to the secretary of state as aforesaid. It shall be the duty of the secretary of state, when such report or reports are delivered to him, to cause 500 copies of the same to be printed and bound in pamphlet form, and, upon the organization of the eighth general assembly, the secretary of state shall distribute to each member thereof three copies of each of said reports, and the remainder of said 500 copies shall be turned over to the state engineer, to be by him distributed to such persons and corporations as are interested in the subject-matter thereof. L. '89, p. 468, sec. 4.

Section 2506. Pay of commissioners. Each of said commissioners shall be entitled to receive for his services, upon the making of said report as herein provided, said report to be accepted by the secretary of state, the sum of two thousand dollars, to be paid upon vouchers approved by the secretary of state, by warrant drawn by the auditor upon the state treasurer. L. '89, p. 468, sec. 5.

Section 2507. Appropriation for expenses. For the purpose of paying salaries and expenses authorized by this act, there is hereby appropriated out of the general fund of the state, not otherwise appropriated, the sum of ten thousand dollars, or so much thereof as may be necessary. L. '89, p. 269, sec. 6.

Section 3657. Board may sell lands to parties constructing ditch—Bond. For the purpose of furnishing irrigation for state lands the state board of land commissioners are hereby authorized, when in their judgment the interest of the state may be subserved thereby, to sell at public sale, at such place as the board may fix, at not less than the appraised value thereof, which in no case shall be less than the minimum price of two dollars and fifty cents ($2.50) per acre, any tract of arid land belonging to the state (except sections sixteen and thirty-six); *Provided*, That not more than one-half section of land shall be sold, and in alternate half sections, to any responsible person or persons, on condition that said person or persons construct an irrigating ditch in such locality and of sufficient capacity to furnish water for the entire tract, and so located that said tract may be irrigated therefrom; *Provided*, That before any of the state lands shall be offered for sale the party desiring to purchase said lands and construct a ditch shall enter into a contract with the board guaranteeing to bid at least the minimum price per acre, and to complete such ditch within a given time, which time shall be fixed by the board in the contract. The contract shall further provide that the party constructing such ditch shall furnish water for the remaining one-half of the state lands at such reasonable rates as the board and the parties building such ditch or canal may agree upon. Such contract shall be drawn by the attorney general and signed by the governor and register of the board and by the party desiring to construct such ditch; *And provided further*, That if any person other than the person making application for the purchase of said lands shall be the highest bidder at the public sale thereof, such bidder shall, within such reasonable time as the board may fix, enter into a contract and bond, as required by the provisions of this act, for the construction of said ditch and for the furnishing of water therefrom; and in the event of his failure to furnish a satisfactory bond and enter into the said contract within the time fixed, then such bid shall be disregarded and such public sale shall be void and of no effect. The board shall make the sale upon like conditions as other state lands are sold, and shall require a good and sufficient bond from the party desiring to construct such ditch, conditioned for the faithful performance of the contract and the conditions of the sale. And in no case shall the title to any of said lands pass from the state until the ditch shall have been completed and accepted by the board. L. '89, pp. 381, 382, sec. 1.

1. This law is, in substance, L. '81, p. 226, sec. 8; repealed by sec. 3653.

Section 3766. **Property exempt from** taxation. The following classes of property shall be exempt from taxation, to-wit: First, mines and mining claims bearing gold, silver and other precious metals (except the net proceeds and surface improvements thereof), for the period of ten years from the first day of July, A. D. 1876; second, ditches, canals and flumes owned and used by individuals or corporations for irrigating lands, owned by such individuals or corporations or the individual members thereof, shall not be separately taxed so long as they shall be owned and used exclusively for such purpose; third, the property, real and personal, of the state, counties, cities, towns, and other municipal corporations, and public libraries; fourth, lots, with the buildings thereon, if said buildings are used solely and exclusively for religious worship, for schools, or for strictly charitable purposes; also, cemeteries not used or held for private or corporate profit.   G. L. '77, p. 742, sec. 2244; G. S. '83, p. 821, sec. 2815.

EXEMPTION.

1.   Ditches free from taxation, sec. 2397; see also Const. Colo., art. X., sec. 3.

2.   Cemeteries exempt, sec. 654; see also Const. Colo., art. X., sec. 5.

3.   Mines heretofore exempt by the constitution, taxed, sec. 3222; see Const. Colo., art. X., sec. 3, note 3.

4.   United States property exempt, sec. 4570.

5.   When the legal title is in the United States and beneficial interest is in the occupant, the state can tax the lands.   County Commrs. v. Cen. Col. Impl. Co., 2 Colo., 635 (1875), reversed 95 U. S., 265 (1877).   When the legal title is wholly in the United States it is exempt from taxation.   Commrs. v. Cen. Colo. Impl. Co., 2 Colo., 636 (1875).

6.   There is no provision constitutional, or statutory which in terms requires the levy of a tax upon the annual net proceeds of mines and mining claims bearing precious metals.   Stanley v. Little P. Mg. Co., 6 Colo., 417 (1882); see Mills' Const. Ann., sec. 438, notes 55-59 and 91.

7.   This section reiterates the permissive language of the constitution in respect to the taxation of the net proceeds of mines, but does not in terms require the levy of the tax.   Id., see Mills, Const. Ann., sec. 438, notes 55-59 and 91.

8.   Articles of commerce in transition are exempt, but sleeping cars leased by a railroad company for its use, though used only in passing through the state, are

not articles of commerce so as to be exempt from taxation. Carlisle v. Pullman P. C. Co., 8 Colo., 324 (1885).

9. The legislature had power under art. X., sec. 3 of the constitution to impose a tax upon mines and all mining property. People v. Henderson, 12 Colo., 371 (1888); see Mills' Const. Ann., sec. 438, notes 78, etc.

10. An exemption of a seminary from taxation by special charter by the legislature becomes a part of the contract, and cannot be impaired. County Commrs. v. Colorado Seminary, 12 Colo., 499 (1889); see Mills' Const. Ann., sec. 438, notes 108, etc.

Section 4403. Paragraph 58. Mills — Ditches— Feeders. To authorize the construction of mills and mill races, irrigating or mining ditches and feeders, on, through or across the streets of the city or town, at such places and under such restrictions as they shall deem proper.

DIVISION XV.

IRRIGATION.

Section 4539. May lease or purchase canal—On vote of electors. Any incorporated town or city in this state shall have power to purchase or lease any canal or ditch already constructed, or which may hereafter be constructed, and all the rights, privileges, franchises of any person or persons, or corporations owning the same, or having any interest or right therein, and to hold and operate the same in the same manner as the persons or corporation from whom the same may be purchased or leased might otherwise do; *Provided*, Such purchase or lease shall be made for the purpose of supplying, by said ditch or canal, water for the use of the people of said city or town; and *Provided further*, That a majority of the qualified electors of such city or town, who shall vote at any regular election which may be held for the election of town officers, shall vote in favor of said purchase. L. '79, p. 198, sec. 1; G. S. '83, pp. 1000, 1001, sec. 3417.

1.　See also as to right of way for water, ditches, etc., and issuing bonds for same.　L. '74, p. 298, etc.

2.　This act is still in force unless repealed by section 4532.

3.　See Irrigation, sec. 2256, etc.

Section 4540.　Shall assume all obligations of owner—Repair—Management.　Any town or city making such purchase or lease shall thereby assume all obligations and other duties which by law devolve upon the owner or owners of such ditch or canal of whom the same may be purchased or leased by virtue of this act; and shall have the power to repair, improve or enlarge the same, or any flume, dam or gate connected therewith; and for such objects may levy and collect taxes in the same manner as other taxes are levied and collected by law.　The management of such ditch or canal shall be under the control of the board of trustees, or council, as the case may be, of such town or city.　L. '79, pp. 198, 199, sec. 2; G. S. '83, p. 1001, sec. 3418.

1.　See section 4403, subdivisions 68, 69.

2.　The care required of a municipal corporation undertaking to supply to residents water for irrigation and using its street gutters for that purpose is such as a man of average prudence and intelligence would employ under like circumstances to protect his own property.　City of Boulder v. Fowler, 11 Colo., 398, (1888).

CORPORATIONS.

(S. B. 53.)

AN ACT

TO PROVIDE FOR THE AMENDMENT OF ARTICLES OF INCORPORATION
OF CORPORATIONS ORGANIZED UNDER THE LAWS OF COLORADO,
AND TO REPEAL AN ACT ENTITLED "AN ACT TO PROVIDE FOR
THE AMENDMENT OF ARTICLES OF INCORPORATION OF INCOR-
PORATED COMPANIES," EXCEPT RAILROAD COMPANIES, AP-
PROVED MARCH 25, 1885, AND ALL ACTS IN CONFLICT WITH THE
PROVISIONS HEREOF.

Be it enacted by the General Assembly of the State of Colorado:

SECTION 1. That any corporation organized under
the laws of this state may amend its articles of incorporation in any respect; *Provided*, No corporation shall, by
amendments, so change its articles as to work a change
in the object or purpose for which such corporation was
originally organized; *Provided*, That any ditch company
may amend its articles so as to allow it to take stock in
telephone companies, for the purpose of affording facilities to such ditch companies in carrying on their business only.

Approved April 6, 1891.

CORPORATIONS—DITCH AND RESERVOIR COM-
PANIES.

(H. B. 98.)

AN ACT

TO ENABLE IRRIGATION DITCH COMPANIES AND RESERVOIR COM-
PANIES TO EXTEND THE TERM OF TIME OF THEIR INCORPORA-
TION.

Be it enacted by the General Assembly of the State of Colorado:

SECTION 1. When the term of years for which
any corporation which has been or may hereafter be,
incorporated as a ditch company, for the purpose of car-

rying water for irrigation purposes, or as a reservoir
company for the storage of water for irrigation purposes,
has expired, or is about to expire by lawful limitation,
and such corporation has not been administered upon as
an expired corporation or gone into liquidation and settlement and division of its affairs, it may have its terms
of incorporation extended and continued, the same as if
originally incorporated, as hereinafter provided.

SEC. 2.   Whenever the corporate life of any such
ditch or reservoir company has expired, or is about to
expire, as aforesaid, the stockholders of such company
may vote upon the question of extending the life of such
corporation for another twenty years, or less, by first
giving notice of such intention by publication, for two
successive weeks, in the newspaper printed nearest the
place where the principal operations of said company
are carried on.   Such notice shall be signed by stockholders owning at least ten per cent. of the entire capital stock of said company, and shall state the place
where and the time when the question of renewal will
be submitted to the votes of the stockholders of said
company, at the meeting held in pursuance of such
notice, provided a majority of the stock of the corporation be represented.   The votes shall be taken by bal-
lot, and each stockholder shall be entitled to as many
votes as he owns shares of stock in said company or
holds proxies therefor; and, if a majority of the votes
cast shall be in favor of a renewal of the corporation,
the president and secretary of said company shall,
under the corporate seal of said company, certify the
fact, and shall make as many certificates as may be
necessary, so as to file one in the office of the recorder
of deeds in each county where they may do business
and one in the office of the secretary of state; and thereupon the corporate life of said company shall be
renewed for another term of not exceeding twenty (20)
years, upon filing the declaration aforesaid, and all
stockholders shall have the same rights in the renewed
corporation as they had in the company as originally
formed.

SEC. 3.   Whereas, in the opinion of the General
Assembly an emergency exists; therefore, this act

shall take effect and be in force from and after its passage.

Approved March 19, 1891.

CORPORATIONS.—DITCH COMPANIES.

(S. C. 331)

AN ACT

TO AMEND SECTIONS 72, 73 AND 102 OF CHAPTER XIX., THE SAME BE-
ING GENERAL SECTIONS 308, 309 AND 338 OF THE GENERAL STA-
TUTES OF THE STATE OF COLORADO, ENTITLED "CORPORA-
TIONS."

Be it enacted by the General Assembly of the State of Colorado:

SECTION 1. Section 72 of chapter XIX., being general section 308 of the General Statutes of the State of Colorado, is hereby amended so as to read as follows, viz:

308. SEC. 72. When any three or more persons associate under the provisions of this chapter to form a corporation for the purpose of constructing a ditch, reservoir, pipe-line or any thereof, for the purpose of conveying water from any natural or artificial stream, channel or source whatever to any mines, mills or lands, or storing the same, they shall in their certificate, in addition to the matters required in section 2 of this chapter, specify as follows, viz: the stream, channel or source from which the water is to be taken, the point or place at or near which the water is to be taken out, the locations as near as may be of any reservoir intended to be constructed, the line as near as may be of any ditch or pipe-line intended to be constructed, and the use to which the water is intended to be applied.

SEC. 2. Section 73 chapter XIX., being general section 309 of the General Statutes of the State of Colorado, is hereby amended so as to read as follows, viz:

309. Section 73. Any ditch, reservoir or pipe line company formed under the provisions of this chapter, shall have the right of way over the line named

in the certificate, and shall also have the right to run water from the stream, channel or water source, whether natural or artificial named in the certificate through its ditch or pipe line, and store the same in any reservoir of the company when not needed for immediate use; *Provided*, That the line proposed does not interfere with any other ditch, pipe-line or reservoir, having prior rights, except |the right to cross by pipe or flume; nor shall the water of any stream, channel or other water course, whether natural or artificial, be diverted from its original channel or course, to the detriment of any person or persons having priority of right thereto, but this shall not be construed to prevent the appropriation and use of any water not theretofore utilized and applied to beneficial uses.

SEC. 3.   Section 102 of chapter XIX., being general section 338 of the General Statutes of the State of Colorado, is hereby amended so as to read as follows, viz: 338.   Section 102.   If any corporation formed for purpose of constructing a road, ditch, reservoir, pipe-line, bridge, ferry, tunnel, telegraph line or railroad line, shall be unable to agree with the owner, for the purchase of any real estate or right of way or easement, or other right necessary or required for the purpose of any such corporation, for transacting its business, or for the right of way or any lawful purpose connected with the operations of the company, such corporation may acquire title to such real estate or right of way, or easement or other right, in the manner provided by law, for the condemnation of real estate, or right of way, and any ditch, reservoir or pipe-line company man in the same manner condemn and acquire the right to take and use any water not previously appropriated.

Approved April 9, 1891.

[Session Laws of 1891.]

WATER FOR DOMESTIC PURPOSES.

(H. B. 139.)

AN ACT

IN RELATION TO WATER FOR DOMESTIC PURPOSES..

Be it enacted by the General Assembly of the State of Colorado:

SECTION 1. Water claimed and appropriated for domestic purposes shall not be employed or used for irrigation or for application to lands or plants in any manner to any extent whatever; *Provided,* That the provisions of this section shall not prohibit any citizen or town or corporation organized solely for the purpose of supplying water to the inhabitants of such city or town from supplying water thereto for sprinkling streets and and extinguishing fires, or for household purposes.

SEC. 2. Any person claiming the right to divert water for domestic purposes from any natural stream, who shall apply or knowingly permit the water so diverted to be applied for other than domestic purposes, to the injury of any other person entitled to use such water for irrigation, shall be deemed guilty of a misdemeanor, and upon conviction shall pay a fine of not less than fifty dollars and not exceeding two hundred dollars, in the discretion of the court wherein conviction is had. Each day of such improper application of water obtained in the manner aforesaid shall be deemed a separate offense. Justices of the peace in their several precincts shall have jurisdiction of the aforesaid offense, subject to the right of appeal as in cases of assault and battery.

SEC. 3. In consequence of the near approach of the irrigation season and to avoid litigation, it is deemed that an emergency exists; and therefore, this act shall take effect from and after the time of its approval.

Approved April 1, 1891.

[Session Laws of 1893.]

CHAPTER 84.

EMINENT DOMAIN.—DRAINAGE OF SURPLUS WATER.

(S. B. 103, by Senator Boyd.)

AN ACT

TO PROVIDE FOR THE DRAINAGE OF WET LAND.

Be it enacted by the General Assembly of the State of Colorado:

SECTION 1. That whenever the owner or owners of any parcel or parcels of land desire to construct a drain for the purpose of carrying off surplus water, and they cannot agree among themselves or with the parties who own the land below, through which it is expedient to carry the drain in order to reach a natural waterway, then proceedings may be had in the same manner as in cases of eminent domain affecting irrigation works of diversion, and the right of way for such drains shall be regarded as equal to that of irrigation canals.

Approved March 21, 1893.

CHAPTER 97.

GAME.

(H. B. 366, by Mr. Dake.)

AN ACT

TO PROHIBIT AND REGULATE THE KILLING, TRAPPING OR OTHER-
WISE TAKING, THE TRANSPORTATION OR SALE OF CERTAIN
ANIMALS, FISH AND BIRDS; TO PROVIDE PENALTIES FOR THE
VIOLATION OF THIS ACT, AND TO REPEAL ALL ACTS OR PARTS
OF ACTS INCONSISTENT HEREWITH.

Be it enacted by the General Assembly of the State of Colorado:

Page 278—Sec. 15. No person shall kill, wound,
ensnare or entrap any beaver within the state of Colorado at any time: *Provided*, That this shall not prohibit
owners of canals or ditches from killing beaver that
interfere with said canals or ditches.

Approved April 7, 1893.

CHAPTER 135.

PUBLIC LANDS.—BOARD OF LAND COMMISSIONERS
—REGULATE DISTRIBUTION OF WATER FROM
STATE CANALS AND RESERVOIRS AND
CHARGE FOR CARRIAGE.

H. B. 511, by Mr. Gordon.)

AN ACT

TO DIRECT THE STATE BOARD OF LAND COMMISSIONERS TO REGU-
LATE THE DISTRIBUTION OF WATER FROM STATE CANALS AND
RESERVOIRS.

Be it enacted by the General Assembly of the State of Colorado:

SECTION 1. Until otherwise authorized by law,
the board of land commissioners is hereby directed to
regulate the distribution of water from state canals and

reservoirs under such rules and regulations as said board shall deem to be for the best interests of the state "and to charge and collect rental for the carriage of water therein."

Approved April 10, 1893.

CHAPTER 107.

IRRIGATION.—Conveyance of Water Rights.

(H. B. 211, Mr. Fitzgerald.)

AN ACT

RELATING TO THE CONVEYANCE OF WATER RIGHTS.

Be it enacted by the General Assembly of the State of Colorado:

SECTION 1. In the conveyance of water rights hereafter made in this state, in all cases except where the ownership of stock in ditch companies or other companies constitute the ownership of a water right, the same formalities shall be observed and complied with as in the conveyance of real estate.

Approved April 7, 1893.

CHAPTER 108.

IRRIGATION—FLOW OF WATER IN DITCHES.

(H. B. 70, by Mr. Crowley.)

AN ACT

TO AMEND SECTION ONE **OF AN** ACT ENTITLED "AN ACT REGULAT-
ING **THE** DISTRIBUTION OF WATER, THE SUPERINTENDENCE **OF**
CANALS OR DITCHES USED FOR THE PURPOSES **OF** IRRIGATION,
AND PROVIDING **A** PENALTY FOR THE VIOLATION THEREOF,
APPROVED **MARCH** 19, 1887.

Be it enacted by the General Assembly of the State of Colorado:

SECTION I. That **section one (1)** of an act entitled
"An act regulating **the distribution of** water, the superintendence of canals **or ditches used** for the purposes of
irrigation, and providing **a penalty for** the violation
thereof," approved March 19, 1887, **be** and the same is
amended to read as follows:

SEC. I. Every person or company owning or **con-**
trolling **any** canal or ditch used for the purposes of irrigation and carrying water for pay shall, when demanded
by the **users** during the time from April 1 until November 1, in **each year, keep a** flow of water therein, so far
as may **be** reasonably practicable for the purpose of irrigation, sufficient **to meet the** requirements of all such
persons as are properly entitled **to the** use of water
therefrom, to the **extent,** if necessary, **to** which such
persons may be **entitled to water, and no** more; *Pro-*
vided, however, That **whenever the rivers** or public
streams or sources from **which the water is** obtained are
not sufficiently free from ice, **or the** volume of water
therein **is** too low and inadequate for that purpose, then
such canal **or** ditch shall be kept with **as** full a flow of
water therein **as** may be practicable, subject, however,
to the rights of priorities from the streams or other
sources, as provided by law, and the necessity of cleaning, repairing **and** maintaining the same in good condition.

SEC. 2. In the opinion of the General Assembly an emergency exists; therefore, this act shall take effect and be in force from and after its passage.

Approved March 25, 1893.

CHAPTER 109.

IRRIGATION.—WATER DISTRICTS NOS. 12 AND 13.

(H. B. 327, by Mr. Wells.)

AN ACT

TO AMEND AN ACT ENTITLED " AN ACT TO REPEAL SECTION FIVE (5) OF AN ACT ENTITLED 'AN ACT TO PROVIDE FOR THE APPOINT-MENT OF A STATE ENGINEER, AND TO DEFINE HIS DUTIES AND REGULATE HIS PAY, AND FOR THE APPOINTMENT OF HIS ASSIST-ANTS AND THE ESTABLISHMENT OF WATER DIVISIONS, APPROVED MARCH 5, 1881, THE SAME BEING SECTION EIGHTEEN HUNDRED AND SIX OF THE GENERAL STATUTES, 1883. AND ALSO TO AMEND SECTION FIFTEEN (15) OF AN ACT ENTITLED AN ACT TO REGU-LATE THE USE OF WATER FOR IRRIGATION, AND PROVIDING FOR SETTLING THE PRIORITY OF RIGHT THERETO, AND FOR THE PAY-MENT OF THE EXPENSES THEREOF, AND FOR THE PAYMENT OF ALL COSTS AND EXPENSES INCIDENT TO SAID REGULATION OF USE.' APPROVED FEBRUARY 19, 1879; THE SAME BEING SECTION SEVENTEEN HUNDRED AND FIFTY-ONE OF THE GENERAL STAT-UTES OF 1883. AND TO ESTABLISH THE SAN JUAN WATER DIVISION, ALSO TO CREATE WATER DISTRICTS IN ESTABLISHED WATER DIVISIONS; ALSO TO PROVIDE FOR UTILIZING TESTIMONY HERE-TOFORE OFFERED AS EVIDENCE IN THE ADJUDICATION OF WATER RIGHTS." APPROVED APRIL 1, 1885.

Be it enacted by the General Assembly of the State of Colorado:

SECTION 1. That section 5 of an act entitled "an act to repeal section five (5) of an act entitled 'an act to provide for the appointment of a state engineer and to define his duties and regulate his pay and for the ap-pointment of his assistants, and the establishment of water divisions.' Approved March 5, 1881. The same being section eighteen hundred and six of the General Statutes 1883; and also to amend section fifteen (15) of an act entitled 'an act to regulate the use of water for irrigation, and providing for selling the priorities of rights thereto, and for the payment of the expenses thereof, and for the payment of all costs and expenses incident to said regulation of use, approved February 19, 1879, the same being section seventeen hundred and fifty-one of the General Statutes of 1883; and to establish the San Juan water division; also ·to create water districts in established water divisions; also to provide for utilizing testimony heretofore offered as evidence in the adjudication of water rights;" approved April 1, 1885, be and the same is hereby amended so as to read as follows:

SEC. 5.   That district number twelve (12) shall consist of all lands irrigated from ditches or canals taking water from that part of the Arkansas river lying in Fremont county, also all lands irrigated from ditches or canals taking water from the tributaries of said portion of the Arkansas river, except that part of Grape creek which lies above the south line of said Fremont county.

SEC. 2.   That section 6 of said act be amended to read as follows: Sec. 6.   That district number thirteen (13) shall consist of all lands irrigated from ditches or canals taking water from that part of Grape creek and its tributaries lying in Custer county.

Approved April 8, 1893.

CHAPTER 116.

LIENS—FOR CLEANING AND REPAIRING UNINCORPORATED IRRIGATING DITCHES.

(H. B. 100, by Mr. Dean.)

AN ACT

TO SECURE LIENS UPON INTERESTS IN UNINCORPORATED IRRIGATING DITCHES OF CO-OWNERS WHO FAIL AND REFUSE TO ASSIST IN CLEANING AND REPAIRING SUCH DITCHES.

Be it enacted by the General Assembly of the State of Colorado:

SECTION 1.   All co-owners of unincorporated irrigating ditches shall pay for the necessary cleaning and repairing of such ditches in the proportion that their

respective interests bear to the total expenses incurred
in said cleaning and repairing; *Provided*, That any
such co-owner may perform labor in cleaning and repairing such ditch, equivalent in value to his or their share
of such expenses as aforesaid , *Provided*, No co-owner
shall be held liable for cleaning and repairing any ditch
below the point from which he takes his portion of the
water.

SEC. 2.   Upon the failure of any one or more of
several co-owners, upon written request of the owners
of one-third ($\frac{1}{3}$) of the carrying capacity, or board of
directors to assist in cleaning and repairing such ditch,
the other co-owner or co-owners shall proceed to clean
and repair the same, and shall keep an accurate account
of the cost and expenses incurred ; and shall, upon the
completion of such work deliver to each of such delinquent co-owners, his agent, lessee or legal representative,
an itemized statement of such cost and expenses.

SEC. 3.   The co-owner or co-owners of any such
ditch or canal who shall clean and repair the same, as
specified in section two (2) of this act, shall have a lien
upon the interest in such ditch owned by such delinquent co-owner for his proportion of such cost and
expense.

SEC. 4.   Any person wishing to avail himself of
the provisions of this act shall file for record in the
office of the recorder of the county wherein the ditch
to be affected by the lien is situated, within thirty (30)
days after the completion of such work, a statement
addressed to the owner or owners of the interest upon
which such lien is claimed, specifying the name of the
ditch and the extent of the interest in the same upon
which such lien is claimed, the date upon which the
work was commenced and the date upon which it was
completed, the total amount expended on such ditch,
and the amount due from such delinquent co-owners.
Said statement shall be signed and verified upon oath
by a claimant.

SEC. 5.   Any party claiming a lien under the provisions of this act may assign in writing his claim and

time to any person, who shall thereafter have all the right and remedies of the assignor.

SEC. 6. No lien claim by virtue of this act shall hold the property longer than six (6) months after filing the statement described in section four (4), unless an action to be commenced within that time to enforce the same.

SEC. 7. Actions to enforce liens claimed by virtue of this act shall be commenced and prosecuted in accordance with the procedure in other civil actions in the state of Colorado. Each party who shall establish his claim under this act shall have a judgment against the party personally liable to him, for the full amount of his claim so established, and shall have a lien decreed and determined upon the ditch interest to which his lien shall have attached, to the extent of his said claims; *Provided, however,* That no judgment shall exceed the interest of the party in such ditch ; nor shall execution issue against other than this said interest in said ditch.

SEC. 8. The court shall cause such ditch interest to be sold in satisfaction of said lien and costs, as in case of foreclosure of mortgages, and in manner and form provided for sales on executions issued out of courts of record, and the owner and creditors shall have a right to redeem, as is provided for in cases of sales of real estate on execution.

SEC. 9. In all actions brought to enforce liens claimed under the provisions of this act in which the plaintiff is successful, a reasonable attorney's fee, to be fixed by the court, shall be assessed against the defendant and shall be taxed as costs in the case. And the plaintiff, if successful, shall also recover all other costs and expenses incured [incurred] in claiming and enforcing his lien.

SEC. 10. The claimant of any such lien, the statement of which has been recorded as aforesaid, on the payment of the amount claimed, together with the costs of making and recording such statement and costs of satisfaction, shall, at the request of any person interested in the ditch interest charged therewith, enter or cause to be entered of record satisfaction of the same ; and, if he shall neglect or refuse to do so within ten (10) days after such request he shall forfeit and pay to the person making such request the sum of ten (10) dollars for every day of such neglect or refusal, to be recovered in the same manner as other debts. Any such statement may be cancelled on the margin of the record by an acknowledgment of satisfaction over the signature of the claimant, or an agent authorized thereto in writing.

Approved April 8, 1893.

CHAPTER 152.

STATE CANAL NO. 1.—BOARD OF CONTROL.

(H. B. 235, by Mr. Gordon.)

AN ACT

CREATING A BOARD OF CONTROL FOR THE COMPLETION AND CON-
STRUCTION OF STATE CANAL NUMBER ONE AND RESERVOIRS
CONNECTED THEREWITH, AND PROVIDING FOR THE CONSTRUC-
TION, COMPLETION, OPERATION AND MAINTENANCE OF THE
SAME.

Be it enacted by the General Assembly of the State of Colorado:

SECTION 1. There is hereby created a board to be known as " The Board of Control of State Canal Number One and reservoirs connected therewith." The said board shall be composed of the lieutenant-governor, who shall be chairman, the state engineer and the warden of the penitentiary. The secretary of the state board of land commissioners shall be secretary of said board of control. Said board is hereby charged with the duty of securing the early completion of state canal number one and reservoirs connected therewith, and of the operation and maintenance of the same as herein provided.

SEC. 2. It shall be the duty of the state engineer to prepare plans and specifications upon and according to the survey already established by the state engineer, and now on file in his office, pertaining to the construction and completion of state canal number one and reservoirs connected therewith. From a point where line of said canal crosses east line of township eighteen, east of and below the present constructed part of the same, east of the "Prison Hogback" near Canon City, south of range seventy (70) west, eastward to the end of said canal as now projected and surveyed; and pertinent to the construction of all appurtenances necessary to the successful operation of said canal, except lateral ditches · therefrom and the head-gates for such lateral ditches. It shall also be the duty of the state engineer to make any survey which may be necessary to the thorough preparation of such plans and specifications, and to submit such plans and specifications as soon as may be to the board of control of state canal number one and reservoirs connected therewith, created and constituted as herein provided.

SEC. 3. Upon the completion and submission of the plans and specifications, as hereinbefore provided, it shall be the duty of such board of control to carefully review and consider the same, and after making changes and modifications therein, if such should seem advisable, to approve such plans and specifications. Thereupon it shall be the duty of the secretary of said board of control to advertise, for at least thirty consecutive days, for bids for the completion of said canal in accordance with such plans and specifications, such advertisements to be made in three or more papers, one or more of which shall be published in the city of Denver, one in the county of Fremont and one in the city of Pueblo. The state board of land commissioners, in order to facilitate the construction of said canal and reservoirs, may, in conjunction with the board of control, advertise and offer for sale, at not less than minimum price, every alternate quarter section of the state and school lands lying under the said canal.

SEC. 4. The board of control may enter into a contract as herein set forth; *Provided, however*, That such contract, if awarded, shall be awarded to the lowest responsible bidder; *And provided, also*, That the said board may, at its discretion, reject any and all bids, in which event they shall direct the secretary of said board to re-advertise for bids for the performance of such work.

SEC. 5. The contract for the completion of state canal number one and reservoirs connected therewith shall be prepared by the attorney general, and shall provide, among other thing, for the completion of such canal in accordance with the plans and specifications approved by the said board of control, within a period of eighteen (18) months from date of said contract; and shall provide for the payment of such work in accordance with the provisions concerning payments hereinafter made in this act. The attorney general shall also prepare a bond in the penal sum of one hundred thousand dollars ($100,000) for the faithful performance of the work of completing said canal in accordance with said contract, which bond must be signed by at least ten good and sufficient sureties satisfactory to a majority of said board of control.

SEC. 6. Payment to the contractors for the material and labor furnished and performed in the construction and completion of said canal shall be made by means of certificates of indebtedness upon the completion of each five (5) miles of said canal, in sums not to exceed ninety per cent. of the state engineer's estimated cost therefor; and upon completion and acceptance by the board of control of the whole line of said canal the whole amount of the contract price shall be so paid in certificates as herein provided.

SEC. 7. Upon the presentation to the auditor of the written acceptance of such work by the engineer, approved as herein provided, it shall be the duty of the auditor to issue in favor of the contractor performing the work, or those duly authorized by him to receive the same, certificates of indebtedness, which certificates of indebtedness shall show upon their face that they bear five per cent. interest per annum, payable semi-annually from the date of their issuance until paid; that they are issued in lieu of immediate money compensation for materials and labor furnished and performed in constructing the state canal number one and reservoirs connected therewith, and such other things and materials as to the auditor may seem pertinent and useful. The certificates of indebtednesss, so issued, shall be for one thousand dollars ($1,000) each, numbered consecutively. One-fourth of the entire amount shall be due five years from the date thereof; one-fourth shall be due and payable ten years from the date thereof; one-fourth shall be due and payable fifteen years from the date thereof, and one-fourth shall be due and payable twenty years from the date thereof. Such certificates of indebtedness shall be countersigned by the treasurer and approved by the governor. These certificates may be accepted by the state in payment for the carriage of water or in payment for lands, and the same shall not in any event become a claim against the state except as to said funds so to be received.

SEC. 8. Upon completion of said canal and its acceptance and approval, as hereinbefore provided, the said board of control of state canal number one and reservoirs connected therewith, shall turn over the said canal, together with all drawings, specifications, reports and records pertaining to said canal, and the action of said board of control, to the state board of land commissioners; whereupon the state board of land commissioners shall assume control of said canal, and shall thereinafter control, operate and maintain the same, subject to such provisions of law as may hereafter be made and established.

SEC. 9. It shall be the duty of the state board of land commissioners to cause the waters carried in the state canal No. 1 and reservoirs connected therewith to be applied to the irrigation of the state lands and all other lands lying under said canal at the earliest convenient and practicable times, and as a means among others to effect such use of water, the board of land commissioners are authorized to offer numerous portions of said lands for lease at such reasonable prices and for such periods, not exceeding twenty years, as will be conducive to the rapid settlement of such lands and the early use of such waters.

SEC. 10. The said board of control of state canal No. 1 and reservoirs connected therewith is here given all the rights and powers that an individual or corporation now has or may hereafter have under the laws of the state, or of the United States, to acquire the right of way over, upon and to any lands necessary for it to use or occupy in the construction and maintenance of such canal.

SEC. 11. It shall be the duty of the state board of land commissioners to establish from time to time reasonable annual charges for the carriage of water, or sell perpetual rights of water, if deemed by it more expedient.

SEC. 12. The title to the said canal shall vest and remain with the state of Colorado, and any money received for the carriage of water therein shall be devoted to the maintenance and operation of such canal; and surplus over and above the cost of operation and maintaining such canal shall be converted into the state treasury, and applied by the state treasurer to meeting the certificates of indebtedness herein provided for and interest thereon.

SEC. 13. It shall be the duty of the said board of land commissioners to construct from time to time, and as rapidly as may seem to such board advisable, lateral ditches and the necessary appurtenances thereto for supplying the lands of the state lying under said canal with water for irrigation, and to see that all of such lands belonging to the state are brought under cultivation within a reasonable time.

SEC. 14. Any receipts or certificates heretofore issued in return for subscriptions and advancement of money by persons owning laug (land) along the line of state canal No. 1 and reservoirs connected therewith shall be received in lieu of money for the lawful and reasonable charges for the carriage of water in the said canal, and all of the certificates hereafter issued, as herein provided, shall be received in lieu of money for charges for the carriage of water in said canal or for perpetual water rights thereunder

SEC. 15. The members of the state board of control of state canal No. 1 and reservoirs connected therewith shall be entitled to their reasonable traveling expenses while performing the duties herein laid upon them, for which amounts the auditor shall draw warrants upon the state treasurer, when such amounts shall be duly certified to him by the secretary of the said board of control.

SEC. 16. All acts and parts of acts inconsistent with the provisions of this act are hereby repealed.

SEC. 17. It is the opinion of the assembly that an emergency exists; therefore, this act shall be in force on and after its passage.

Approved April 17, 1893.

CHAPTER 153.

STATE RESERVOIR.—MONUMENT CREEK APPROPRIATION.

(H. B. 166, by Mr. Wootton by request.)

AN ACT

TO PROVIDE FOR THE STATE RESERVOIRS AT MONUMENT CREEK, IN EL PASO COUNTY, AND FOR THE COMPLETION OF SAID RESERVOIR AND APPROPRIATING MONEY FOR THE PAYMENT OF THE SAME.

Be it enacted by the General Assembly of the State of Colorado:

SECTION 1. There is hereby appropriated out of any money in the state treasury belonging to the internal improvement fund, and not otherwise appropriated, the sum of four thousand dollars ($4,000), or so much thereof as may be necessary for the purchase of lands for the state reservoir at Monument Creek, in El Paso county, and for the rip-rapping or paving the spill way to said reservoir.

SEC. 2. The board of construction for said reservoir as named in the act providing for the construction

therefor approved April 16, 1891, is hereby authorized
and directed to purchase and acquire title, in behalf of
the state, to the land actually to be covered at high
water mark by the waters in said reservoir and by the
dam and also a strip fifty feet in width extending entirely around said reservoir, pond and dam; *Provided*,
That no greater sum per acre shall be paid for any part
of the lands thus purchased than that the board of ap-
praisers has already set upon the same.

SEC. 3.   The said board of construction is further
authorized and directed to have the spillway or waste
weir of the said reservoir rip-rapped or paved for greater
security, this work not having been included in the
original estimate or contract.

SEC. 4.   The auditor of state is hereby authorized
to draw warrants for the payment or the purchase of
said lands and for the paving or rip-rapping of said spillway upon vouchers certified to by the aforesaid board, not
exceeding the sum of four thousand dollars ($4,000(;
Provided, That if there is no cash in the treasury belonging to said fund, then the state auditor is hereby
authorized to sell any warrants or the securities in his
hands belonging to said fund; *Provided*, Such securities
shall not be sold for less than par and accrued interest
thereon.

SEC. 5.   In the opinion of the General Assembly
an emergency exists, therefore this act shall take effect
and shall be in force from and after its passage.

(NOTE.—This bill was filed with the secretary of
state without the govervor's signature and without his
objection on May 3, 1893, and hence became a law
under constitution, art. V., sec. 2.   Secretary of state.)

INDEX.

INDEX.

www.ingramcontent.com/pod-product-compliance
Lightning Source LLC
Chambersburg PA
CBHW021709210326
41599CB00013B/1579